PREHISTORIC BRITAIN

The Stonehenge Enigma

Robert John Langdon

ABC Publishing Group

Published by ABC Publishing Group

12-14 High Street

Rottingdean, Brighton

East Sussex BN2 7HR

www.abc-publishing-group.co.uk

SECOND EDITION Version 2.1

Dedicated to the three R's

Alexandra, Jack and James, whose 'encouragement'

lead me to the discovery of Stonehenge's true construction date

in the summer of 2009

ISBN: 978-1-907979-04-0

Edited by Mike Davis and Annette Holliday

CONTENTS

This book is the conclusion to a chain of events,
that started with a course in Archaeology at the
Museum of London in the 1990s
and ended in a thunderstorm at Stonehenge in August 2009.

17000 BCE	10000 BCE
Ice Age Ends	Mesolithic Age Begins

Preference to the Second edition

The publication of the first edition of this book was undertaken under some haste as considered opinion suggested that such 'ground breaking' evidence about one of the world's most famous monuments, would cause a massive stir that the demand for further information would be overwhelming. So it would be wise to publish what we had to protect our copyright and then amend thereafter.

What we did not expect was the total dismissal of the hypothesis and the refusal to take the matter seriously. Mainstream organisations and archaeologists ignored the publication or attempted to discredit the subject matter without even reading the book. Even so, the book sold in the thousands and it generated good publicity with those who appreciated to read and understand the new information.

The facts in the book remain the same and the evidence clearly shows that Stonehenge was first built five thousand years before commonly accepted dates. It's a sad indictment that such discovers in the past have suffered the same academic dismissal such as the discovery of Doggerland at the start of the last century by Clement Reid in 1913.

The archaeological establishment and most archaeologists dismissed his claims as pure fiction although Fishing boats and trawlers had constantly dredged up animal bones and stone tools from the bottom of the North Sea, for over 100 years. It was only by chance that gas and oil reserves was found in this area of the world that large corporations started to scan the sea bed and drill pilot holes in their search for wealth, that this hypothesis was taken seriously and in 1990 – 75 years after Clement first published his book 'Antiquity of Man' that Prof Bryony Coles reinvigorated the subject and termed the name 'doggerland' for the region – by then Clement and his sceptics were long dead.

This is a poignant reminder that discoveries that change the world's view of history are not easily accepted by the establishment and particularly academics are judged against their past published literature, therefore their sceptical attitude towards my book was understandable.

Therefore, the second edition has had the time and maturity to dot the i's and cross the t's and add even further information and evidence, which I believe now to be beyond refute.

To this end, I have completely rewritten the first section on Geological Evidence, with extracts of new evidence that have come to light, including references from recently published books from recognised experts that now support by original hypothesis. My new location in the South Downs as also gave me a great opportunity to study the 'dry river valleys' in complete profile (rather than just borehole data) as they cut through the chalky cliffs, which are exposed to the sea. Not only can you see the significant colour and material change of the water that was once present, but

timeline				
4500 BCE	**2500** BCE	**800** BCE	**0-400** AD	**1-2000** AD
Neolithic Age Begins	Bronze Age Begins	Iron Age Begins	Roman Period	Written History

you can touch and see the enormous turmoil the chalk and flint experienced, showing the aquatic power and extent of these greatly misunderstood features.

I have also decided to add the Stonehenge construction data which I was going to save for one of my future books, as the other parts of the trilogy, that make up Prehistoric Britain, have grown to such huge proportions in word and page size that it seemed sensible to place this 'landmark' information in this book, so that it is not out of context.   This will include an illustration of how Stonehenge really did look once the large Sarsen stones were introduced, and the extraordinary mathematics our ancestors used for its construction.

Yet I can assure you, that the 'truth is still out there' and is ready to be discovered by persons with an 'open mind' and a sound judgement, which my dad used to call 'common sense'. Enjoy the book and for those how find enlightenment, as Pink Floyd once said, then 'I'll see you on the dark side of the moon'.

RJL

Rottingdean 2012

I love prehistory; I think it should be called pre-mystery. In my mind, it's the greatest ever 'who done it'. Agatha Christie or even Dan Brown would have been proud of leaving so many tantalising clues and artefacts about what happened so long ago in mankind's ancient past.

So who am I then? Holmes, Poirot or Indiana Jones?

Well hopefully a combination of all three as I love to solve puzzles, and this book answers the most captivating of all questions – who built Stonehenge and why? Just think of the clues on offer: strange stone monuments; relics of a bygone age; scientific evidence that seems to contradict each part of the puzzle as it's discovered and an overwhelming realisation that this is not a game - this is reality!

I want to solve the mysteries from the dawn of our civilisation. If that fails to excite you to the bone, I guess nothing in history ever will.

To understand our ancestry, you must be able to detach your mind from the 21st century. You need to picture the land that archaeologists call the 'stone age period' - the problem is that your mind has already created a mental picture of either hairy fur covered men dragging their women into the cave for fun or Fred Flintstone and Barney Rubble going for a drive in their stone-mobile. Of course, neither of these images is correct or helpful.

As you read this book you will journey with me back in history. You will need to remember that for a considerable time, the people you will read about, may have only possessed wood and flint as tools yet

they still had the foresight, capability, tenacity and organisational skills to build monuments that would last 10,000 years. I really don't think even our best known recently built structures – the O2 Millennium Dome, Wembley Stadium, Canary Wharf Tower - will survive for one tenth of that time period. I would strongly argue that we must give our ancestors the respect they deserve and be proud that our forefathers created such a great civilisation.

The story starts with a car journey, driving home from a family holiday in the summer of 2009. I had previously studied Archaeology at the Museum of London in the early 1990s. During my course the standard format of materials was poorly photocopied archaeological texts and illustrations - the use of an available overhead projector was far too modern for this type of dusty establishment.

The illustrations provided showed that mankind originally moved from Africa to the Middle East and then finally, onwards to Northern Europe and eventually Britain. This was the sacred proven pathway of our civilisation, and if anyone dared to write an essay even hinting at an alternative suggestion,

timeline

4500 BCE	2500 BCE	800 BCE	0-400 AD	1-2000 AD
Neolithic Age Begins	Bronze Age Begins	Iron Age Begins	Roman Period	Written History

they were branded a heretic and would be marked down accordingly – as you may begin to imagine, I fell into this category.

I have always found this 'traditional' model of our civilisation's pathway difficult to accept or understand. In my mind, civilisations are incredibly old and diverse and need many tens of thousands of years to develop the characteristics we see around the world today - therefore to suggest that the first farmers were migrants from Africa who travelled to the Middle East and who then transferred their knowledge onto Europe over only a few thousand years, seems totally unbelievable, naive and somewhat simplistic because in my view, that's not how civilisations develop. I also found it surprising that the literature and teaching provided, failed to include any references to the even more diverse civilisations of the Far East – so, I guess, according to traditional theories provided, the Chinese must have never discovered farming and therefore they must still be living in caves today?

Nonetheless, it is true to say that the antiquated lesson structures and information to which I was exposed did give me an insight into how Archaeology itself has evolved – via a group of amateur enthusiasts, whose dated theories somehow still remain prominent today. Many of these old academics have little to no engineering or practical skills, let alone the empathy to understand the true nature of hunter-gatherers or the issues surrounding self sufficiency in a hostile environment (thinking a little more like Ray Mears would have helped them considerably).

I am often amused by watching archaeologists spending hours on the most boring, labour intensive work such as drawing an excavation plot on an A3 board with strings as guides. On a few occasions, I've approached the poor student conscript allocated this God forsaken task and mentioned the marvellous new invention called a digital camera – which not only takes high resolution shots, but if used to take photographs from different angles and heights, they can then be turned into a 3D map on a PC. The poor students usually look at me as if I'm mad to suggest such a device and some well indoctrinated ones have suggested that "cameras can miss things".

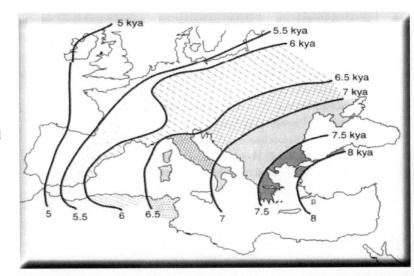

This is a clear indication of how the whole archaeological process has developed - more as a 'club' rather than a progressive scientific subject. Hence the quiet distain showed to the utterly engaging TV series Time Team in which a site is excavated in 3 days. The archaeological establishment sees this as 'popularist' and not 'true archaeology' which, in their view, should take years and eventually reach the same old, tired but established conclusions that are automatically accepted as being correct - and certainly wouldn't allow a theory that would take them out of their academic comfort zone.

17000 BCE	10000 BCE
Ice Age Ends	Mesolithic Age Begins

There are many examples of this - archaeological political correctness - even within the Time Team television programme. When they find something unusual or not easily identifiable, the word 'religious' or 'ceremonial' is suddenly produced as an obvious explanation, rather than a more truthful admission of, 'Tony, I ain't got a clue mate!'

This is what I call 'archaeological self regulation'. It's a way of guaranteeing your future career as an archaeologist.

As will be revealed later in the book, often when archaeologists are given scientific evidence from carbon dating that contradicts the 'traditionally' held view, it's dismissed as an 'anomaly' - this is the usual response when scientific evidence conflicts with the archaeological accepted belief system.

This 'strange' approach to the science of archaeology can also be seen when sites are dated. A majority of sites are dated by broken pottery or flint finds discovered within the site. This tenuous link is based on a premise that ALL pottery and flints can be dated by its design or structure. In some sites, it's absolutely true to say that these type of finds can be a good source of evidence if it is found 'in situ' with other items such as coins or other carbon dated material.

However, to rely on this as a form of evidence when pieces are found on the surface is problematical, as ALL it shows is that this type of pot was used on this site – AT SOME POINT after this type of pot was made – not necessarily at the SAME TIME.

For example, if an archaeologist in the future finds the ruined remains of St. Paul's Cathedral (and imagine that all written records for the site were lost) he or she would attempt to date the site by the artefacts found on site. Say, for instance, a Coca-Cola can crammed into the base of a remaining wall was discovered – using

current accepted practice, the archaeologist would conclude that, because the can was found 'in situ', the construction date of St. Paul's would be approximately 1950 to 2000 – the practice simply doesn't work.

Even a child would realise that dating an item found at a site, doesn't automatically mean that the site shares the same date. Sometimes, complete sites are dated just by fragments of pottery without any other evidence, other than a 'gut feel' or 'tradition'. This can readily be seen by the number of 'iron age' encampments found on OS maps. Most of these sites are on hills and have ditches surrounding them; archaeologists automatically classify these monuments as 'iron age' as the perception is that they are fortifications, built at a time of extreme violence

timeline

4500 BCE	2500 BCE	800 BCE	0-400 AD	1-2000 AD
Neolithic Age Begins	Bronze Age Begins	Iron Age Begins	Roman Period	Written History

– the Iron Age. There is not one piece of true evidence to support this claim, yet we have thousands of monuments incorrectly classified as 'iron age' all over Britain. This type of unsupported evidence would simply not be deemed acceptable by any other 'science' (apart from Geology - and we'll talk about that later).

Frustratingly, however, it is accepted practice in archaeology. So the next time you ask an archaeologist "what's that" and they mention either the 'religious or ceremonial' word, stamp really hard on their foot or ask **"what religious group or ceremony would that be?"**

I witnessed at first-hand, this kind of blanket prejudice when I submitted my final essay during my archaeology course in the 1990s. The essay was about Stonehenge and highlighted the conflicting evidence throughout the site. This particularly applied to the car park post holes that had been ignored by archaeologists when discovered in the 1960's and in so doing, they missed how this find could have helped to establish the true dating of Stonehenge instead of relying on the 'loose' pottery and antler evidence that's currently taken as the absolute truth.

I will never forget the comments my lecturer wrote on my marking sheet, which seem even more poignant today as I write this prologue: "Would make the basis of a good book, but has no credibility for serious archaeology today". I suppose I should have been happy to have received a pass mark, even if it was only just!

Anyway, back to the plot. I was driving to London via the A303 which takes me past Stonehenge, when suddenly, day turned to night and a cold eerie storm ripped across Salisbury Plain. I watched in the slow traffic as the poor tourists, in their summer clothes, ran as best they could for shelter while the traffic crept to a halt. At that point, my mind started to drift, and I looked around at the grassy fields as they started to become waterlogged.

I was driving past a point called 'Stonehenge Bottom', a deep ravine adjacent to Stonehenge. The hills were now feeding water down to the lowest point of the valley and the water was very quickly becoming very deep as it reverted back to being the river it once had been. "You idiot!" I said out loud to no-one in particular.

It was a phrase I had started to use a lot in everyday life, as I had become a great admirer of Hugh Lawrie's 'House' – I could identify with the same

House M.D.

17000 BCE

Ice Age Ends

10000 BCE

Mesolithic Age Begins

stubborn, rebellious and analytical qualities of the TV personality. (If you haven't managed to catch any of these enthralling programmes, I'd highly recommend them).

The reason for my outburst was that I had driven and walked past this same spot more times than I care to remember but I had never realised that this was a huge clue to the 'post hole puzzle' I had considered so many years before, in my essay.

I got off the road and returned to Stonehenge. As I entered the car park, I was guided to the auxiliary car parking spaces on the grass behind the tarmac section. There was some chaos as attendants were busy trying to fence off a large central section of the grassy car park as it had started flooding. Most drivers found this naturally quite annoying, but I had such a huge smile on my face that I'm sure the attendants must have thought I was insane.

But....what if this was wrong?

You see I had been told, as are all archaeologists that have studied the site, that the riverbed where the car park lay was pre-ice age (at least 400,000 years ago if not more) so consequently it's always been ignored by archaeologists.

But....what if this was wrong?

As I stood in the rain watching the river return to Stonehenge, I asked myself 'why is the car park, still flooding?' – If the experts were right, despite this extraordinarily heavy rainfall, should not be flooding as the 'dry river valley' (in which the car park was situated) had supposedly dried up hundreds of thousands of years ago. Only a significant rising of the water table would cause it to flood now – not this relatively small level of rainfall I was witnessing. Yet the evidence I saw with my own eyes told me otherwise. I knew I had to go back and look at the evidence from the start, and this time, I would question everything, not assume that so called 'accepted' theories were correct and would literally leave no stone unturned.

If the experts were wrong, this small piece of the jigsaw would suddenly reveal not only the darkest secrets of Stonehenge, but the true date of the great civilisation that had created the stone monuments of Britain.

timeline

4500 BCE	**2500** BCE	**800** BCE	**0-400** AD	**1-2000** AD
Neolithic Age Begins	Bronze Age Begins	Iron Age Begins	Roman Period	Written History

Introduction

This book has been written to explain and prove a hypothesis that I have been working on over the last 30 years.

In publishing this work I understand that it represents a fundamental change to not only British history, but to the history or the world. Consequently, I have not undertaken this lightly, but there are moments in the evolution of any science – and yes, archaeology is a science - when a new theory will challenge the fundamental beliefs of that science's existing structure and that is the objective of this book.

These progressive challenges should in no way be viewed as criticism of the current theories, but a logical succession, creating a more coherent set of beliefs that moves the science forward and helps everyone involved to develop a greater understanding of the subject.

It allows experts to re-examine the subject matter in a new light and extract the truth from the myth - that has sometimes been responsible for creating false realities.

In my view, Archaeology (and in some respects Geology) has not been challenged enough over the years especially to the same extent as we have seen in other better funded sciences such as Physics or Biology. If this unchallenged acceptance was extended to these and other sciences, we would probably still be living in a world without Quantum Mechanics or Darwin's theory of evolution.

What you will see unfolding in this book is a newer form of archaeology, which I refer to as 'Landscape Dating'. It allows us to date sites, not only from the findings on the site but critically, from their location in the prehistoric landscape. This science will bring a new interpretation and understanding of the structural complexities and the philosophies of our ancestors.

I believe that through 'Landscape Dating' and through the evidence presented in this book, history will no longer paint a tainted portrait of fur- covered hairy men, running half naked, chasing deer and mammoths over Salisbury plain. This image will be replaced by the vision of an idyllic landscape of water and tree lined islands as we currently see in certain locations in Russia and Northern Canada, within which an intelligent and sophisticated civilisation existed - a society with advanced engineering skills, living a pleasant serene Mediterranean sailing existence, in perfect harmony with their fellow man and nature.

Furthermore, this unique civilisation went on to travel the world trading, teaching and living with the local populations ultimately sharing their engineering and philosophical knowledge with the Ancient Greeks and Egyptians, amongst others

I consider myself a prehistorian and philosopher with a flair for 'Landscape Dating'. For I must confess that I was never one for getting down and dirty or sifting through thousands of fragments trying to find a decent piece of pottery with which to identify or date a site. Although this form of 'treasure hunting' does give its participants the thrill of discovery, for me, the larger picture of trading, politics and alliances made these discoveries interesting.

17000 BCE	*10000* BCE
Ice Age Ends	Mesolithic Age Begins

History of any site can best be seen within the landscape and at the location of the magnificent monuments. When you see henges or stone circles on the edge of cliffs or peninsulas, the power and awe of these ancient monuments is absolutely breathtaking. So when the final pieces of this ancient mystery eventually fell together I felt I had no option but to place my life on hold and write this book.

But in the process of collating my findings and thoughts, the overriding evidence became so massive that our single book became a complete trilogy that eventually traced the roots of the Stonehenge builders.

As I have examined these ancient monuments in depth, I have considered them from the perspective of the engineer and social philosopher within me rather than as an archaeologist and that attitude and style is represented in the context this book. I have tried to lay out my hypothesis in a jargon free, logical and sensible way with evidence that I hope you will consider being sufficient proof to enable you to reconsider what you currently believe to be accepted history. In some instances, I will turn accepted theories on their head, as the evidence can be interpreted in a completely different way to the current accepted theories.

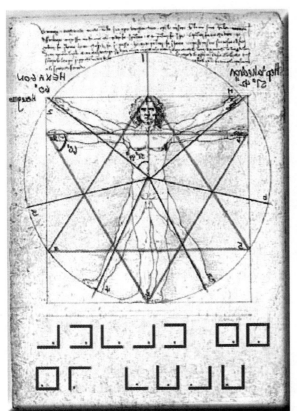

Other Archaeology books I have read in the past have frustrated and confused me. They either ramble on about the author's friends and lunches they have had while researching the book or focus on irrelevant issues. I will do neither in this book – I will simply try to present 'the facts' and my views based on these facts in 'layman's terms' and an informative, interesting manner. It's only by engaging in the evidence that future debates can be progressed and by so doing, our understanding enhanced.

What we must bear in mind is that the evidence is 'absolute' but the interpretation of this evidence is open to scrutiny and debate. Therefore, to assist clarity, I have laid out my hypothesis (in full) in a clear and concisely manner - this is what the next chapter is dedicated to. This will allow you to understand what I am trying to prove.

I will then go through the evidence to date, based on the four major sites that surround and include Stonehenge – Woodhenge (Durrington Walls), Avebury and Old Sarum. This will allow you the reader to make your own mind up on what is the truth and what is not possible.

I have always admired writers and, during my recent university courses, presenters who can take complex subjects and turn them into simple analogies which allow anyone to understand the concepts without reverting to jargon or technical references. If you have studied Quantum Mechanics

timeline

4500 BCE	2500 BCE	800 BCE	0-400 AD	1-2000 AD
Neolithic Age Begins	Bronze Age Begins	Iron Age Begins	Roman Period	Written History

and Philosophy as I have, you quickly realise which lecturers really do understand their subject and which are just treading water.

Indeed, one of my favourite film scenes is from 'Philadelphia' when Denzel Washington turns to Tom Hanks, who is in the process of explaining a complex legal problem and asks him to **"tell it to me as if I was a six year old"**. A sure way of saying, give it to me straight and simple! Well, I hope this book doesn't quite talk to you as if you're a six year old, but I will explain some of the complexities and mysteries of Archaeology in a down to earth, clear and precise manner, using similar analogies when necessary.

Once we have exhausted the topic or subject manner, Einstein will appear to conclude the debate and clarify what I term as 'proof of my hypotheses'. They will be forty or so proofs which are the basis of the evidence in the formulation of this book; a complete list will be shown in Appendix A with reference to the pages where the evidence is gathered. This enables me to claim that my book is not just a handful of 'ideas' like most hypothesis, but a collection of evidence that proves the hypothesis beyond doubt – even if I do get one or two wrong. This is quite possible as we are flying at the edge of understanding and it is therefore possible that we will interpret evidence incorrectly due to either bad reporting or fieldwork which is beyond my control.

A large part of this book is my landscape surveys of Mesolithic and Neolithic Sites in the Stonehenge area. This and other 'case studies' are central to the book as I see them as the 'best evidence' for my hypothesis. Too many good ideas look feasible on paper but when they are studied in detail they are shown to be 'just ideas' without true substance or methodology.

I could have chosen any area in Britain to prove my theory, but Stonehenge provided me with more archaeological evidence than any other as it has been the centre of prehistoric interest since the Roman invasion 2,000 years ago. It also has the most detailed analysis of any site in Britain as money has in this location has been well spent unlike most other sites - and that is still not enough.

What we have found in Stonehenge is a direct connection to at least three other main sites in the same area, which makes it the centre or hub of Prehistoric Britain, a little like London today.

"tell it to me as if I was a six year old"

17000 BCE	10000 BCE
Ice Age Ends	Mesolithic Age Begins

Later books will explain in detail why this is an ancient centre of activity, but the book contains sufficient information to prove that Stonehenge is not what it seems and it was built at a time unknown to present archaeologists.

It has always been recognised that post-glacial landscape is still a mystery to both geologists and archaeologist alike, as quoted in the book 'Stonehenge in the Landscape' by Michael Allen "In short, we are dealing with a period from the upper Palaeolithic to the Late Neolithic covering nearly five millennia for which, realistically, we know little from the environment except by assumption and inference from the adjacent area way from the chalk"

To give you a flavour of how the other books of this trilogy will unfold, the final chapter unites the evidence of the former chapters creating an unique insight and vision into what kind of great civilisation must have existed in prehistoric times that could organise and build such colossal ancient monuments that have lasted over ten thousand years. For archaeologists have always sadly failed to understand the type of culture that was required to socially organise and manage large numbers to create these types of structures.

This 'lost civilisation' has enormous consequences for the history of not only Britain but also Europe and the World. We will show you that this civilisation clearly uses (for its time) engineering skills and mathematics way beyond what archaeologists and historians to date given credit for their abilities.

Enjoy the book and I hope it will encourage you to go out and re-explore these ancient sites armed with a new vision and knowledge of how our landscape really used to look some 10,000 years ago. For each unexplored prehistoric site still holds great secrets of our 'lost civilisation' past just waiting to be rediscovered. Those who dare to venture forth 'with an open mind' could be actively participating in solving the disentanglement of the greatest historical mystery of all time.

timeline

4500 BCE	2500 BCE	800 BCE	0-400 AD	1-2000 AD
Neolithic Age Begins	Bronze Age Begins	Iron Age Begins	Roman Period	Written History

The Hypothesis

12,000 years ago the last Ice Age finally melted, revealing the Britain we know today – or did it? Britain had been underneath two miles of ice and the surrounding seas had frozen solid. What was left was a huge icy mass of enormous weight, pushing down on this tiny island. This mass had compressed the earth so much that the land surface lay at least a half a mile below the sea bed as we know it today.

So what exactly happened after the great ice age melted?

This huge mass of watery ice that covered Britain raised the groundwater tables and left the land totally saturated. In fact, the volume of water was so great that it eventually created the English Channel and the North & Irish Seas. Very slowly, the land then started to rebound, so slowly in fact that even now, parts of Britain are still rising about one cm per year. This melting ice combined with the lowering of the land levels created not the single island called Britain we know today, but a series of smaller islands and waterways - totally unrecognisable to the landscape that is so familiar to us now.

Moreover, the land became a sub-tropical forest as the warmer climate that first melted the polar ice caps encouraged the growth of abundant foliage. This would very probably have caused the islands to experience what we would consider today to be monsoon seasons. Which in turn then kept groundwater tables abnormally high for another 4,000 years? The foliage, groundwater and warm climate would have left the islands resembling the Amazon's rain forests, rather than the grassy hills of Britain we see today.

The only way our Mesolithic ancestors of Britain could have had to adjust to this new environment would have been to develop and use their boat and seamanship skills. Consequently, adapting to living and trade by these shorelines travelling via the vast waterways and lakes rather than through the forests, which would have been riddled with dangers such as brown bears, packs of wolves and wild boar all roaming freely. Therefore, the shorelines became critical – our Mesolithic ancestors would have lived, worked and gathered by these 'super-waterways' and would have created social monuments and beacons on their beaches and peninsulas.

My hypothesis proposes that our greatest prehistoric monuments, such as Stonehenge, were built on these watery peninsulas. I also propose that the ditches surrounding henges were NOT dry ditches, as archaeologists currently believe, but were, in fact, constructed to be watery moats and canals, which turned these sites into very special islands. The most astonishing aspect of this hypothesis is that even today, thousands of years after the groundwater have subsided, we can re-visit these sites and identify the ports and channels of this bygone age and by using the landscape, we can date, more accurately than ever before, when the sites were initially constructed.

17000 BCE	*10000* BCE
Ice Age Ends	Mesolithic Age Begins

Another key component of my hypothesis is the discovery of navigational signposts, built within the landscape, on the banks of these waterways. Traditionally, archaeologists have believed that these had religious or ceremonial use, whereas my evidence shows that they had a more functional engineering purpose, helping our ancestors to navigate around these islands.

In fact, the book will show for the first time, that these signposts guided our ancestors when they transported enormous stones from the Preseli Mountains in Wales to the site at Stonehenge – by boat. These very stones were used to build the first phase of this magnificent monument. Even more importantly, I can show the exact location of where these vast stones were unloaded from the boats and how the mechanism by which this precious cargo was lifted onto the shore.

This discovery of this landing site has led us to accurately date for the first time the original construction date of Stonehenge - 7500BCE to 8000BCE. This is 5,000 years earlier than current archaeological estimations, making Stonehenge the oldest Monument in the world.

timeline

4500 BCE	2500 BCE	800 BCE	0-400 AD	1-2000 AD
Neolithic Age Begins	Bronze Age Begins	Iron Age Begins	Roman Period	Written History

Section One – The GEOLOGICAL EVIDENCE

In this section, we will look at the sequence of events that occurred directly after the last Ice Age and the consequences it had on the environment and landscape of Britain. If my hypothesis is correct, it will prove that these events increased the groundwater tables, not only in the South of Britain, but more importantly, at the case study site, Stonehenge.

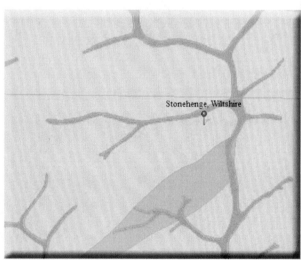

But this has not always been the case

If you study any British Geological Society (BGS) map of Britain you will notice it shows a series of bedrock, sedimentary and superficial deposits. At a scale of 1:50,000 km and below these deposits start to form a labyrinth of material that look like canals and gigantic waterways which lay under the surface on top of the bed rock. This lays testament to how the landscape must have looked at some stage of our natural history and this is particularly prevalent in the Stonehenge area as well as other chalk bedrock outcrops.

These superficial deposits that resemble ancient rivers can clearly be seen on the surface and are known to the archaeologists, geologists and the general public as 'Dry River Valleys' - because the river valleys are currently dry.

But this was not always the case!

Until recently geologists believed that the contours of these chalk hills and valleys were cut during a 'Periglacial Phase' of the 'Quaternary Period', which is the current geological period that started about 2.6 million years ago - although there is no real evidence of the exact date of their formation. Recent theories (and in Geology these new ideas are occurring on a regular basis) suggest that these dry river valleys are the result of water flooding, washing away the top soils and rounding the chalk sub-soil, during the melting period after an Ice Age.

The problem for archaeologists and geologists is - which one? - For there were several during the quaternary period.

Geologists seem content to give rough estimations on the construction date of geological objects such as dry river valleys, which for the archaeologist can become misleading. For the origin of these objects is of some interest, the actual dates when they could have been used by man is even more important if we

Stonehenge, Wiltshire

BGS superficial deposits around Stonehenge

17000 *BCE*

10000 *BCE*

Ice Age Ends

Mesolithic Age Begins

are to understand the anthropological implications and, through this process, any archaeological findings in relation to their locations.

So we must best try to understand not WHEN the dry river valleys were formed, but when LAST did they have water running within them?

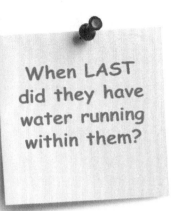

When LAST did they have water running within them?

Geological maps clearly indicate that great rivers once flowed through Britain and we know that the greatest deluge of water that has ever affected the landscape is at the end of an ice age, when the gigantic ice caps finally melt, at the end of the last ice age some 17,000 years ago, geologists have estimated that the ice was over two miles thick in some places. This substantial level of ice MUST have created huge flooding all over the Mesolithic landscape including the Valleys of the South Downs, even thought they were over 100 miles away from the main ice sheet.

Modern geologists now accept that the dry river valleys are the product of water (not ice as previously believed) and looking at some extreme examples of the soil erosion and valleys cut, we are not talking about just frozen tundra slowly melting in the summer seasons - but millions of gallons of fast flowing water cutting away at the top-soil and sedimentary deposits, all the way down to the bedrock in some instances.

This geological evidence can clearly be seen in the cliffs and valleys of the South Downs. Just like the Stonehenge region, this area has the same chalk sedimentary bedrock and ancient post-glacial rivers. Evidence for these rivers are found by the subsoil consisting of sand, silt and clay. This subsoil can be seen in the valleys (known as deans) of the South Downs and most graphically in the exposed face of the white chalky cliffs that have been eroded by the sea giving us a perfect 'dissection' of a typical prehistoric waterway.

The South Downs showing the sand by the top-soil

Modern geologists have yet to identify these huge concave sections of the cliffs, as being the remains of the ice melt from the last glaciation which had filled with water leaving the sandy sediments embedded in the chalky sedimentary rock face, just after the great melt, some 15,000 years ago, instead they claim they are 'windblown' loess or wash from the valley walls.

What they can't explain is the relatively short distance from the sandy soil to today's top soil and the exact date of this sandy sediment. If you look closely at the cliffs, you will see the sandy remains of the river is touching the top soil. If this dry river valley was as old as some archaeologists and geologists suggest - where is the rest of the top soil?

timeline

4500 BCE	2500 BCE	800 BCE	0-400 AD	1-2000 AD
Neolithic Age Begins	Bronze Age Begins	Iron Age Begins	Roman Period	Written History

If the top soil erodes as quickly as some 'experts' also suggest - why is there 18 inches of top soil on top of the chalk today?

There should be none or are we expecting some massive climatic event to wipe away the top soil in the near future, or is the dating of the prehistoric river beds and consequently dry river valleys totally incorrect?

As a matter of practice, archaeologists investigating Stonehenge have always ignored the obvious dry river valleys that surround the site. For they have incorrectly perceived that this area looked similar as today in the time of Stonehenge's construction and therefore, they falsely believe that these river valleys were dry not wet in the Mesolithic and Neolithic Periods.

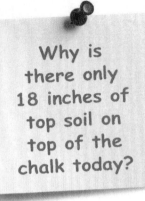

Why is there only 18 inches of top soil on top of the chalk today?

Proof of Hypothesis No. 1

Due to the size and location of the last glaciation, water from the ice cap must have flooded the landscape causing the rivers to rise and turn the landscape into a 'flooded island' environment

17000 *BCE*   **10000** *BCE*

Ice Age Ends   Mesolithic Age Begins

Chapter 1 – The Land called DOGGER

One of the strangest mysteries in archaeology is the disappearance of Doggerland. This was a land mass the size of Ireland, which lay approximately 50 miles east of Norwich, between Britain and Denmark - within what is now known today as the North Sea.

Until very recently, we were unaware of Doggerlands existence. Past geological theories proposed that the North Sea and the English Channel were formed by the rising water at the end of the last Ice Age 15,000 years ago. This theory was accepted for a long period of time as it fitted with the traditional view of our prehistory. However, over the past 100 years fishing boats and in recent times Dutch trawler men, while fishing at the bottom of the North Sea, repeatedly found evidence of tools from post-Ice Age occupation.

Initially, because of existing archaeological theories, it was suggested that these finds were from a pre-Ice Age Europe, when Britain was still connected to the continent. Subsequent evaluation, however, showed that the level of sophistication in the construction of these tools that showed to be the work of skilled Mesolithic or Neolithic people, i.e. after the Ice Age.

Sonar images obtained by the wealthy companies prospecting for oil and gas provided the answer to this mystery by discovering a land mass we now call 'Doggerland' which lays 15 to 60 metres under the North Sea. The current view held by both geologists and archaeologists is that this land did in fact existed in Mesolithic period (10,000 BCE to 4500 BCE) but disappeared with the rising sea levels, caused by the melting ice caps.

However, this theory is fundamentally flawed.

The problem is the Ice Age ended at least 5,000 years BEFORE the Doggerland was swamped - geologists will tell you that the water came from ice cap of the last ice age. But it does not take 5,000 to 10,000 years for an ice cap to melt in the Northern Hemisphere. If it did, why is there so much fuss about Global Warming today – if the risk of flooding was 5,000 years down the line? The truth is that there is a large hole in the current post Ice Age theory.

Where was the water for those thousands of years before it engulfed Doggerland?

In an attempt to validate this unfounded, almost ludicrous sea level theory, they have even suggested that tsunamis played a part in the submergence of Doggerland. Now, it is quite possible that Doggerland was hit by a massive wave resulting from the collapse of a continental shelf, but as we have seen recently with the 2004 Indian Ocean disaster in Sumatra and more recently 2011 in Japan, the wave only temporarily covers the land, it does not sink the landmass completely.

After the initial destruction, the water would then recede. At the time geologists estimate this tsunami, 7,000 years ago, it would not have covered Doggerland

However, this theory is fundamentally flawed.

4500 BCE	2500 BCE	800 BCE	0-400 AD	1-2000 AD
Neolithic Age Begins	Bronze Age Begins	Iron Age Begins	Roman Period	Written History

permanently unless the sea level had risen at the same time and obviously, vast quantities of NEW water would be needed for that event. The second problem with the ice cap melting theory is one of displacement. When ice melts in the sea, it does NOT raise the sea level as reported in our media. This simple process can easily be seen with ice cubes in a glass of water. When the ice cubes melt the water in the glass does not overflow, in fact the water level decreases, because (as any housewife knows) when you freeze water in your fridge the volume increases. Consequently, frozen ice displaces more water not less!

So why did the sea level rise and cover Doggerland?

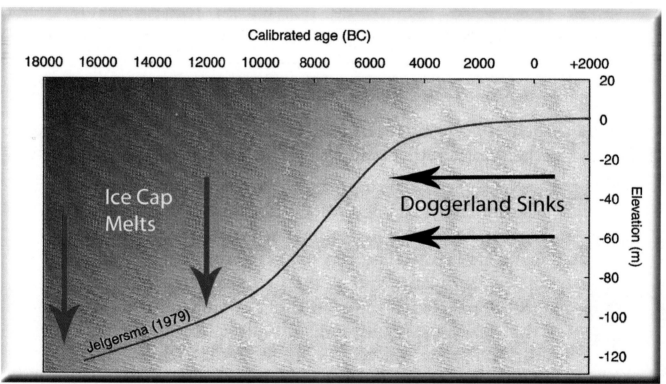

Doggerland sinks thousands of years after the ice melts

17000 *BCE*   You are **HERE**   **10000** *BCE*

Ice Age Ends   Mesolithic Age Begins

The simple answer is that the water that makes up the North Sea and English Channel was "sitting on" the land and was released into the sea very slowly, over a period of time. My hypothesis being proposed here is that, when the ice melted on the land, the environment was swamped by immense floods and as we have now shown, these floods can be seen in the geological post-glacial river bed as superficial deposits of sand and gravel. All of the melting ice water did not run into the sea for of two main reasons; the land mass had been compressed (known to geologists as an 'isostatic transformation') and therefore sunk into the earth's surface due to the colossal weight of the ice cap ('glacial loading') and secondly, the nature of the soil below us allows water to seep into the bedrock which is stored in vast groundwater reservoirs known as 'aquifers'.

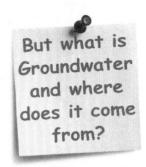

Over the course of the next 10,000 years the groundwater slowly seeped from the aquifers through the rock bed into the sea. This was because the land mass started to recover from the ice age squashing and started to rise up to its normal height "Isostatic Rebound". This released greater amounts of groundwater into the sea, to eventually cover the low lying lands including Doggerland.

This recovery from the 'isostatic transformation' is still seen in parts of Scotland as the landmass around the Forth, Tay and Clyde Valleys rises about 2mm every year. All this evidence suggests that at the end of the ice age the British landmass flooded with so much groundwater that it took up to 10,000 years for all this flood water to move from the soil and bedrock to the seas - raising sea levels.

But what is Groundwater and where does it come from?

Hydrology and Groundwater (from the UK Groundwater Forum)

The presence and flow of groundwater is the key element to my hypothesis, so it is worth spending bit of time in understanding what this substance is and how it flows invisibly under our environment.

When a hole is dug in permeable rocks, at a particular depth water begins to flow in. The surface of the water that accumulates in the hole is the water table and the water in the ground below the water table is groundwater. The variations in the shape of the water table reflect the topography in a subdued form. The water table is near the ground in valleys, actually intersecting the ground surface where rivers, lakes and marshes occur, but it is at much greater depths below hills.

The pore spaces of rocks are saturated with water below the water table and groundwater is said to occur in the saturated zone. Immediately above the water table, water is drawn up into pore spaces by capillary forces into a thin zone called the capillary fringe. Rocks above the water table, including the capillary fringe, form the unsaturated zone; although they do contain water they are generally not completely saturated, and the water cannot be abstracted.

timeline

4500 BCE	*2500* BCE	*800* BCE	*0-400* AD	*1-2000* AD
Neolithic Age Begins	Bronze Age Begins	Iron Age Begins	Roman Period	Written History

Water is continually moving through the environment – we call this the water cycle. Water evaporates from the oceans, condenses into clouds and then falls on the land surface as rain, only to flow into rivers and back into the sea. However, there is one aspect of the water cycle that is often forgotten – groundwater. Rainfall doesn't only reach rivers by running off over the land surface.

Most of the rainfall will soak into the soil, which acts like a giant sponge. In the soil some of the water will be taken up by plants and, through a process called transpiration, will return to the atmosphere, but some will soak further into the ground – a process called infiltration - and trickle downwards into the rocks, becoming groundwater. The level at which the rock becomes saturated is called the groundwater table. Water in this saturated zone will flow from where it has infiltrated to a point of discharge. This might be a spring, a river or the sea. Much of the flow of a river will be made up of discharging groundwater, and groundwater provides a vital role supporting wetlands and stream flows.

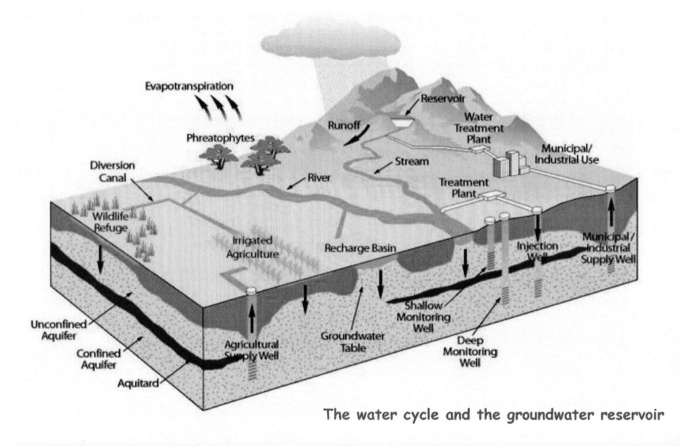

The water cycle and the groundwater reservoir

17000 BCE   You are HERE   **10000** BCE

Ice Age Ends   Mesolithic Age Begins

Water is present almost everywhere underground, but some geological formations are impermeable – meaning that water can hardly flow through them – and some are permeable – they contain fine holes that allow water to flow. Permeable formations that contain groundwater are known as aquifers. The holes that water flows through can be spaces between individual grains in a rock like sandstone, or they can be networks of fine cracks. Very occasionally groundwater will flow in underground rivers, but this is the exception rather than the rule.

Groundwater comes from rain. The average annual rainfall over the UK is about 1100 millimetres, ranging from more than 2500 millimetres over highland Britain to less than 600 millimetres on the lowlands of eastern England. A significant part, almost 500 millimetres in lowland areas, evaporates, mainly in the summer. The remainder is available to infiltrate permeable rocks although where the rocks have low permeability or, where they are overlain by layers of relatively impermeable clay, part will flow over the ground as surface runoff. Water infiltrates the ground mainly in the winter and slowly moves down through the unsaturated zone, eventually reaching the water table and becoming groundwater.

After temporary storage in the ground, groundwater drains from springs and seepages into streams and rivers. Maximum discharges occur at the end of the winter when groundwater levels are high following the seasonal infiltration. They steadily decline throughout the summer into the autumn. The contribution that groundwater makes to the flow of rivers is called base flow and it is responsible for maintaining the flow of rivers during extended periods of dry weather when surface runoff virtually ceases.

Groundwater provides about one-third of public water supplies in England and Wales, 7% in Northern Ireland and 3% in Scotland. The regional differences reflect the distribution of aquifers and the more favourable conditions for the development of surface water resources in both Northern Ireland and Scotland.

Groundwater makes up nearly 70% of all the worlds freshwater; only 0.2% is found in lakes, streams or rivers and 30% is bound up in snow and ice on mountains and in the Polar Regions. As rivers and lakes tend to be supported by groundwater, it is not exaggerating to say that almost all the water we use for agriculture, industry and drinking water is either groundwater or has been groundwater at some point in the water cycle.

Our major aquifers are the Chalk, Jurassic Limestones and Permo-Triassic sandstones, with substantial supplies available from other formations, such as river gravels and Carboniferous Limestone. Over recent years groundwater have been bottled for sale as mineral waters. Often these waters come from minor aquifers in upland areas where distinctive geology imparts interesting flavours.

It should be remembered that not all groundwater resides at sea level. The natural motion of water means that it moves from rainfall and springs down to rivers and eventually to the sea. However, the pace of this flow back to the sea is dependent on soil types and conditions. This is why we have lakes and reservoirs at high altitudes as they are situated on a form of impervious soil that does not readily leak water and they are constantly fed by small streams and constant rainfall - this variation is known as THE GROUNDWATER TABLE and varies throughout Britain.

timeline

4500 BCE	2500 BCE	800 BCE	0-400 AD	1-2000 AD
Neolithic Age Begins	Bronze Age Begins	Iron Age Begins	Roman Period	Written History

My analogy

This process can be illustrated more clearly if Britain is pictured as a giant dry sponge. If an ice cube was placed on the sponge, over time, the ice would melt and the water would soak into the sponge. The key factor here is the melting into the sponge, not out of the sponge to the surrounding areas. Eventually, the water will seep out of the sponge given sufficient volume leaving the majority of the water still in the sponge, this represents the nature of 'groundwater' and it has kept us alive since man has evolved two million years ago.

Geologists and Archaeologists would have us believe that the water from the last ice age just disappeared into the sea without a massive reaction to the land surface. If all the water from previous ice age just flowed into the sea without a detour into the massive aquifers under the soil then the rivers would have run dry long ago and man would have died for he would not have been able to dig the wells he has always needed to survive, if groundwater did not exist.

Proof of Hypothesis No.2

Water from the ice cap from the last Ice Age flooded the British landscape resulting in newly formed and enlarged rivers with islands – this groundwater slowly receded from the land and moved to the North and Irish Seas, creating the English Channel and flooding Doggerland.

Chapter 2 – The Big Squeeze

As we have clearly shown in the previous chapter, when the ice caps covered the Northern Hemisphere at the end of the last Ice Age, it had a dramatic effect of on our planet's hydrology. The weight of the ice also had another impact on the Continental Crust which scientists refer to as 'Isostatic or Earth Transformation'. This process is so tremendous and widespread that it would have affected the whole of Britain, for thousands of years after the last Ice Age. The extent of this geological compression at our case site, Stonehenge to date is little known, so to understand the true nature of the phenomenon we need to look at more detailed investigations undertaken in other countries, to allow us to estimate the degree of transformation, which could have occurred.

Case Study - 'Shorelines of the North American Great Lakes During the Past 20,000 Years'. James Clark from Wheaton College, 2008.

During historical times, water levels of the modern Great Lakes have fluctuated by more than a metre. This is largely caused by changing weather patterns and the associated rates of evaporation and discharge of rivers and groundwater entering and leaving the lakes. As ice advanced over the Great Lakes region the earth's surface under the ice subsided, while the region beyond the ice bulged upward. In general, as the ice retreated, the process occurred in reverse.

Within Clark's findings, he indicated that the earth's surface had been compressed by as much as 700 metres over the last 20,000 years and has only recently (in geological terms) 'bounced back' to what

Great Lakes showing Depth of land surface over time

we would regard as its present level. It's logical to conclude therefore that here in Britain; a similar kind of transformation must have taken place. Geologists claim that the ice sheet spread down the West coast over the Irish Sea, as far as the Scilly Isles, but no further south than the Thames.

What's critical to consider is the volume/weight of water contained within the Irish Sea, in comparison with that within the Great Lakes. The Great Lakes have a volume of some 23,000 cubic km of water, whereas the Irish Sea is 13,500,000 cubic km of water – over 600 times larger and therefore heavier than the Great Lakes.

Present geological convention suggests that this huge volume of ice only affected the landscape and groundwater tables only in Scotland, Wales and Ireland but not the rest of Britain. Geologists believe this because they have been able to monitor the existing rebound in Britain which shows that the North West of the country is still rising by 2 cm per annum whereas

timeline

4500 BCE	2500 BCE	800 BCE	0-400 AD	1-2000 AD
Neolithic Age Begins	Bronze Age Begins	Iron Age Begins	Roman Period	Written History

the South is sinking by 2mm per annum – the view being that if the South of Britain also had been covered by the ice sheet, it too would be rising.

Their conclusion is that the ice cap did not reach the South East corner of Britain and stopped along the River Thames, across to Bristol. The most overlooked feature of this isostatic rebound process is the fact that once squashed; the land actually rebounds back ABOVE its original height before the process started, before returning to its original position, as shown in the Great Lakes study. The simple reason to this unique effect is that we live on a continental shelf (or crust) that floats on the volcanic magma of the earth's core. We are well aware that continental drift affects everyone as the 'tectonic plates' float on this 'sea of lava' but few geologists have recognised how isostatic transformation affects this mantle. It is my belief placing a large weight on a floating surface (such as the continental crust) and then removing it would induce a diminishing 'sine wave' effect on the land.

If pressure is applied – like a ice cap on the land it will sink in relation to the sea. When released it will spring back, the fact that it is higher only reinforces the view that a see-saw motion will continue as a minor 'after effect', as we see in earthquakes, getting progressively smaller over time.

Is that what we are seeing in the SE corner of Britain?

If we are experiencing a smaller 'after shock' and going down by 1mm per annum, that eventually could stop and we may go up by 0.5mm in the future – this would mean that the isostatic transformation affected all the country directly, which indirectly effect the groundwater levels.

> *Glaciation - is defined as the formation, movement and recession of glaciers. At present, glaciers cover about 10% of the world's land area (14.9 million km2). Most of this is under the Antarctic and Greenland ice sheets; only about 700 000 km2 is covered by the thousands of glaciers in the remainder of the world. Glaciations have been much more extensive in the past than it is today, occurring mostly as large continental ice sheets.*

So what effect does Isostatic transformation have on the groundwater levels?

If Britain 20,000 years ago was under the influence of say 110,000 giga tonnes of ice the land would have been squashed at least 800m (half a kilometre) below its original surface, if we take into account the result of the Great Lakes case study. Even without any additional water, the groundwater table would be on or above surface height, as the ice melted the land would 'rebound back' very, very slowly. As we see in Scotland this rebound is still happening some 10,000 years after the last ice had left.

This ice was replaced by water, not a little bit of water but a huge flood of water, recent studies in the US has shown that this release of water had devastating effects in regions where this water was trapped behind the melting ice and caused massive destruction upon release.

17000 *BCE*   **10000** *BCE*   You are HERE

Ice Age Ends   Mesolithic Age Begins

Case Study - The Rocks Don't Lie: A Geologist Investigates Noah's Flood by David R. Montgomery 2012.

"Geologists long rejected the notion that cataclysmic flood had ever occurred—until one of them found proof of a Noah-like catastrophe in the wildly eroded river valleys of Washington State.

After teaching geology at the University of Washington for a decade, I had become embarrassed that I hadn't yet seen the deep canyons where tremendous Ice Age floods scoured down into solid rock to sculpt the scablands. So I decided to help lead a field trip for students to see the giant erosion scars on the local landforms.

We drove across the Columbia River and continued eastward, dropping into Moses Coulee, a canyon with vertical walls of layered basalt. We gathered the students on a small rise and asked them how the canyon had formed. They immediately ruled out wind and glaciers. The valley was not U-shaped like a typical glacial valley, and none of us could imagine how wind might gouge a canyon out of hard basalt. But neither were there rivers or streams. After a while I pointed out that we were standing on a pile of gravel. I asked how the rounded granite pebbles came to be there when the closest source of granite lay over the horizon. Silence.

Hiking through eastern Washington canyons littered with exotic boulders is a standard field trip for beginning geologists. It takes a while to register what you see. A dry waterfall hundreds of feet high in the middle of the desert. Giant potholes where no river flows today. Granite boulders parked in a basalt canyon. Gradually the contradictions fall into place and a story unfolds. Where did wayward boulders the size of a car or house comes from? What was the source of the water that moved them around and carved the falls? Today, even novice geologists can conjure up eastern Washington's giant floods.

Long before the discovery of the scablands, geologists dismissed the role of catastrophic floods in interpreting European geology. By the end of the 19th century such ideas not only were out of fashion but were geological heresy. When J Harlen Bretz uncovered evidence of giant floods in eastern Washington in the 1920s, it took most of the 20th century for other geologists to believe him. Geologists had so thoroughly vilified the concept of great floods that they could not believe it when somebody actually found evidence of one.

Bretz was a classic field geologist and a controversial figure throughout his career. In 1925 he presented the story of the region's giant floods, seeing what others at first could not—and then would not—see. He

timeline

4500 BCE	2500 BCE	800 BCE	0-400 AD	1-2000 AD
Neolithic Age Begins	Bronze Age Begins	Iron Age Begins	Roman Period	Written History

spent his lifetime piecing together the story of how a raging wall of water hundreds of feet high roared across eastern Washington, carving deep channels before cascading down the Columbia River Gorge as a wall of water high enough to turn Oregon's Willamette Valley into a vast backwater lake.

Bretz found exotic granite boulders perched on basalt cliffs hundreds of feet above the highest recorded river level. In the scablands, a desolate region stripped of soil, he came across dry waterfalls and potholes hundreds of feet above the modern river. Gigantic gravel bars deposited within dry valleys implied deep, fast-flowing water. Streamlined hills rose like islands, extending more than 100 feet above the scoured-out channel ways.

He realized the chaotic landscape had been carved by an enormous flood that chewed deep channels through hundreds of feet of solid basalt. The ancient flood deposited an enormous delta around Portland, Oregon, backing up flow into the Willamette Valley. The waters, he eventually realized, could have come from catastrophic drainage of Lake Missoula, an ancient, glacier-dammed lake in western Montana.

Bretz was ridiculed until 1940, when geologist Joe Pardee described giant ripple marks on the bed of Lake Missoula. The 50-foot-high ripples, he said, were formed by fast-flowing currents and not by the sluggish bottom water of a lake. Only sudden failure of the glacial dam could have released the 2,000-foot-deep lake. The catastrophic release of 600 cubic miles of water through a narrow gap would sweep away everything in its path. In 1979, when Bretz was 97 years old, the Geological Society of America awarded him its highest honor, the Penrose Medal.

Recognition of the Missoula flood helped other geologists identify similar landforms in Asia, Europe, Alaska, and the American Midwest, as well as on Mars. There is now compelling evidence for many gigantic ancient floods where glacial ice dams failed time and again: At the end of the last glaciation, some 10,000 years ago, giant ice-dammed lakes in Eurasia and North America repeatedly produced huge floods. In Siberia, rivers spilled over drainage divides and changed their courses. England's fate as an island was sealed by erosion from glacial floods that carved the English Channel. These were not global deluges as described in the Genesis story of Noah, but were more focused catastrophic floods taking place throughout the world. They likely inspired stories like Noah's in many cultures, passed down through generations.

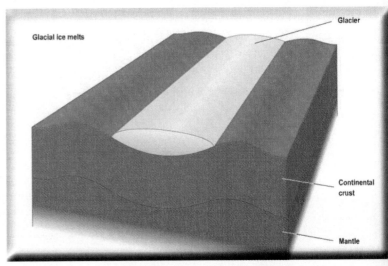

Isostatic Transformation

17000 BCE — 10000 BCE — You are HERE

Ice Age Ends — Mesolithic Age Begins

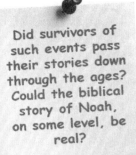

Did survivors of such events pass their stories down through the ages? Could the biblical story of Noah, on some level, be real?

Since devastating floods were a fact of life on the margins of the world's great ice sheets, people in those areas probably witnessed them. Early missionaries in eastern Washington reported stories of a great flood among Yakima and Spokane tribes, who could identify locations where survivors sought refuge. An Ojibwa Indian legend from around Lake Superior tells of a great snow that fell one September at the beginning of time: A bag contained the sun's heat until a mouse nibbled a hole in it. The warmth spilled over, melting the snow and producing a flood that rose above the tops of the highest pines. Everyone drowned except for an old man who drifted about in his canoe rescuing animals. The native inhabitants of the Willamette Valley told stories of a time the valley filled with water, forcing everyone to flee up a mountain before the waters receded."

Did survivors of such events pass their stories down through the ages? Could the biblical story of Noah, on some level, be real?

Case Study - Europe's Lost World

In my search for the builders of Stonehenge, we have looked in-depth to a land just a couple of hundred miles to the east called Doggerland. We have found to our astonishment, when observing the seismic surveying of this area of the North Sea, that the sinking of Doggerland (between 9000 - 4000BCE) gives us a 'blueprint' of not only how Doggerland, but moreover, Britain would have looked directly after the last Ice Age. This is an 'edit' extract for copyright reasons of a passage from the recent book 'Europe's lost world' - the rediscovery of Doggerland by Gaffney et al 2009, who have also studied the same region.

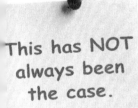

This has NOT always been the case.

If there are problems with interpreting what the Doggerland vegetation looked like in the past, presumably we are on safer ground with respect to the various landscape features identified through the seismic analysis of the North Sea. After all, we know what rivers looked like, don't we?

Anastomosing Rivers

Unfortunately, most people derive their impression of how the river valleys looked in the past from how modern valleys looked today and therein lays the problem. Large modern rivers like the Thames, Trent, Seven and Avon, run through broad alluvial flood plans and the clays in the flood plain are now a major agricultural resource. Somewhere in the vast expanse of cultivated lands that exist in river valleys will actually be a river whose present course only takes up, perhaps 10% of the available land in the flood plain. There is occasional flooding, but generally the river course is stable.

4500 *BCE*	2500 *BCE*	800 *BCE*	0-400 *AD*	**timeline** 1-2000 *AD*
Neolithic Age Begins	Bronze Age Begins	Iron Age Begins	Roman Period	Written History

This has NOT always been the case.

Work on the sedimentary and environmental records of the river valleys of Britain has clearly suggested that our image of a river valley is a recent creation. Modern valleys, particularly those with clay fills seem to date mainly from the Bronze Age or Early Iron Age onwards[1]. They also appear to be a product of the expansion in agriculture that occurred at about this time and the associated increase in soil erosion that led to large amounts of clay and slit entering into the river valleys.

During the Holocene, these rivers would have changed. This is indicated by a number of studies from Britain that clearly show the nature of early Holocene large river systems. Over time, silt would have built up over these valleys. Waterside vegetation and woodland would have developed and stabilised the channel sides, giving some degree of permanence to the course of the rivers.

However, the rivers appear to have maintained a mesh-like appearance, with channels filling the whole of the available floodplain even though these could be several miles across. These channels are relatively unstable and when they failed, they did so spectacularly. Once the bank sides collapsed in a storm or flood, the underlying sands and gravels would shift at speed. Channels would fail and rivers would shift violently across the floodplain.

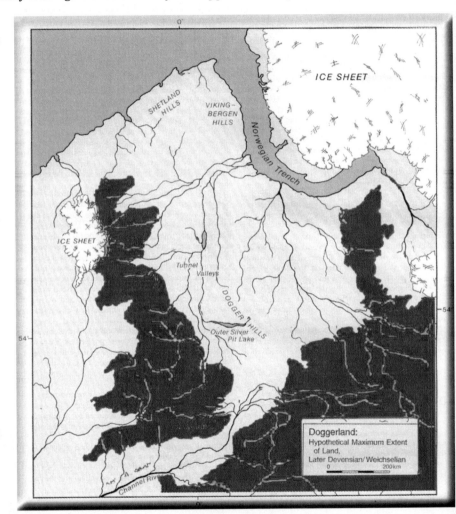

Doggerland:
Hypothetical Maximum Extent of Land,
Later Devensian/Weichselian

0   200km

The result is an 'anatomising' river system that occupies most of the valley floor. This produces a floodplain that bears no resemblance to those of today. Abandoned river channels are overgrown by meadows, grasslands and scrub woodland. Flatlands can be expanded to become large areas of swamp, reed beds and Carr woodland. Gallery woodlands of willow and at a later date, alder dominate the channel sides with more distant and stable areas developing groves of mixed deciduous woodlands. Given the dominance of this kind of river system in the past in Britain, similar environments must be present in the area of the large river systems that dominated both Doggerland and Britain.

Is what we are seeing on the floor of the North Sea a 'blue print' of how Britain as a whole looked in Mesolithic times?

Extensive areas of dense reed bed, wet carr, willow, birch and alder woodland, open pools of water and marshes that attracted man. These 'wet' environments were connected by a labyrinth of rivers and canals so Mesolithic man navigated this land and surrounding lands such as Britain, France and Germany, as the only way you could move from area to area without swimming and leaving your goods behind would be by a boat. But according to present theories, boats will not be used for another 5,000 years.

Proof of Hypothesis No.3

The Isostatic transformation of the landmass during the last ice age that left the landmass nearly half a mile below the sea level would have raised the level of the WATER TABLE relative to the landmass.

timeline

4500 BCE	2500 BCE	800 BCE	0-400 AD	1-2000 AD
Neolithic Age Begins	Bronze Age Begins	Iron Age Begins	Roman Period	Written History

Chapter 3 – WATER, water everywhere....

Before we look at further supporting data for the flood after the Ice Age, we need to see if it is possible that there is other evidence that has either been overlooked or incorrectly dated. To do so, we must understand how geology is mapped and dated.

Geologists have found evidence that the River Avon, which runs near Stonehenge, was much higher in the past than at present. They estimate that in 400,000 BCE the groundwater was 30 – 40m higher than it is today. Our hypothesis indicates that the rise in the groundwater table was due to the Isostatic Transformation after the last Ice Age. We believe these dates currently indicated by geologists are slightly out of sequence.

The Geology of Britain is the study of the composition, structure, physical properties, dynamics, and history of the Earth's materials, including the processes by which they are formed, moved, and changed.

The challenge lies in the fact that geologists are looking at millions of years of history within their data, whereas archaeologists only look at the last 10,000 years. By investigating such a vast expanse of time, geologists find it more difficult to identify all detailed aspects of the last 10,000 years, referred to as the Holocene Period, as it only represents a tiny proportion of the time span they investigate. Another problem is the way geologists date sub-soils and mineral layers.

If we look at the known flood plans of the local river Avon we notice that there is clear evidence that this river has flooded in the past. However, how this evidence is dated is another question.

In a publication 'Crustal uplift in Southern England: evidence from a river terrace records'[2] Illustrated that there were several terraces from the river Avon still visible in the Hampshire basin which they called T5 - T10.

'Little has been done to determine the age of either the terrace sequence or the older River Gravels in the Avon Valley'.

In short **'we don't really know'**, but they go on:

'However, interglacial sediments within the two lowest terraces of the Solent have been assigned to OIS 7 and OIS5e (Allen et al., 1996) Organic Remains of probable Ipswichian(OIS 5e) age (Barber and Brown, 1987) and probable early Devensian age (Green et al.,1983) have been described from sediments underlying low-level terraces within the Avon valley, but at higher elevations in the valleys of the Avon and its tributaries, no organic remains have ever been described'.

17000 *BCE*　　　You are HERE　　　　　**10000** *BCE*

Ice Age Ends　　　　　　　　　　　Mesolithic Age Begins

Bet you
can see the
problem
here!!

They imply that these terraces are between the Ipswich (135,000 BCE to 71,000 BCE) and the Devensian period (71,000 BCE to 12,000 BCE) - clearly Geologists estimate dates with no degrees of relative accuracy, which we try to use in Archaeology. But what is quite remarkable is the 'probable' word - where is the carbon dating? This 'shot in the dark' is confirmed from this extract from the document:

'Palaeolithic artefacts are, however present in the terrace deposits of the Avon at various sites (Blackmore, 1864, 1865, 1867; Reid, 1885; Harding and Bridgland, 1998)....

But hang
on a
minute!

Material in undoubted primary context does not seem to have been described, but the bulk of the archaeology appears to have come from Terrace 7 (Clarke and Green, 1987) for which an age in the middle Pleistocene can, thus assumed...'

So the geologists are dating segments of River Terraces by Archeological finds!

So how are the Archaeologists dating the finds in the first place, well there are two options. Firstly it can be done by the condition of the find; in this case these are Stone Axes. Now to be honest One Stone Axe looks very much like another, depending on the skill level of the 'knapper'. Archaeologists take a guess that the better quality of the knapper the younger the stone tool.

Bet you can see the problem here!!

If the stone knapper was just rubbish, then the tool would be given a date later than the true date of production - that's far too easy to do so archaeologists confirm the dates by where the find was laying in the geological record.

But hang on a minute!

The Geologists rely on the archaeological finds to date the geological segments, which in turn is dated by the archaeologists relying on Geological dating – this is known a a circular argument.

Yes you're right - it's nonsense, as neither the Geologists nor the archaeologists know the date of these flood plans as they rely on each other for the dates. Consequently the flood plans could be any date in the past 70,000 years including the dates just after the last ice age.

What is most interesting is that 90% of all the tools found in the Avon River Terrance investigation were found on Terrace number 7 - the exact same height that matches our hypothesis - a coincidence??

timeline

4500 BCE	2500 BCE	800 BCE	0-400 AD	1-2000 AD
Neolithic Age Begins	Bronze Age Begins	Iron Age Begins	Roman Period	Written History

If our hypothesis is correct not only must the river Avon be much higher in the past than today, but ALL the rivers in Britain during the Mesolithic (10,000 BCE to 5,000 BCE) in PARTICULAR the rivers that are the Avon feeds before reaching the Sea and one of these Rivers is the Thames.

CASE STUDY - "The Holocene Evolution of the River Thames", Jane Sidell, Keith Wilkinson, Museum of London 2000.

In 2000 the Museum of London, published a book based on the research undertaken when the extension to the Jubilee Line was being planned.

This research was written by Jane Sidell, Keith Wilkinson , Robert Scaife and Nigel Cameron, all experts in their field working for either the Museum of London or associated Universities. The book 'The Holocene Evolution of the London Thames', did not raise much interest even though the conclusions should have alerted the archaeological world to the fact that the Holocene (immediately after the ice age) environment was much changed from today.

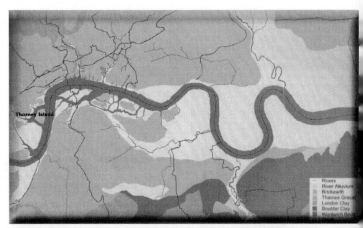

They found that the Thames was 10 times larger in the Mesolithic Period directly after the Ice Age than today and Big Ben and the Houses of Parliament were on an island called 'Thorney'. Their report showed that the Thames was even much bigger just 2000 years ago compared to today when the Romans first discovered our capital city.

This is why the first roads were built via Thorney Island as the City of London was impossible to cross without a boat. Later when the Romans had need of deep water harbour, the city of London was used and bridges were built to span the river.

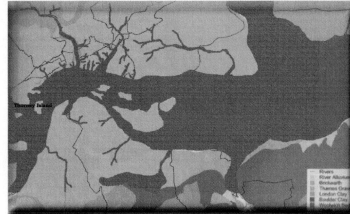

They found masses of 'alluvium' in the mouth of the Thames indicating the size of the river when it was first created. This could be accurately dated as the Thames as we see today did not exist until the end of the last ice age cut the new channel as we have seen from other case histories from America.

River Thames in Roman and Mesolithic Periods

17000 BCE   You are HERE   **10000** BCE

Ice Age Ends   Mesolithic Age Begins

As the River Thames is fresh water and in the Mesolithic Period the Sea water levels were 65m lower than today - where did all that water come from to fill the Thames to that extent?

And the ONLY answer to this question is 'groundwater levels' and rivers that feed the Thames directly. Consequently, these volumes will in turn need to be ten times greater than today. So with all that extra groundwater and swollen rivers - how would Britain look in the Mesolithic Period?

The River Thames is feed by many rivers including the Kennet and River Avon, both of which would needed to be 10 times larger to feed the Thames the water it needed to create the 'alluvium' our archaeologist found in the Lower Thames. This means the River Avon would go from being 65m high at Amesbury to 97m high. At this height Stonehenge would become a peninsula surrounded by groundwater and the Mesolithic post holes, found in 1966, would have been on the shoreline - for they were used to moor the boats that brought the stones from the Preseli Mountains in Wales - a simple and direct route in the Flooded Mesolithic.

The River Thames clearly shows that water directly after the ice age raised groundwater levels in Britain and therefore proves my hypothesis that the landscape was flooded in the early Mesolithic period. Although the evidence from the roman occupation shows that some 10,000 years later the Roman still found the River so high that it took another 400 years before the city of London started to take shape as we know it today.

This slow drop in groundwater levels after the great melt can be seen very graphically in our second case study on the South Downs. In this case study we can see that these high groundwater levels still affected our land even just 500 years ago.

Proof of Hypothesis No.4

The river Thames was ten times larger in the Mesolithic period than today and of 'fresh water' as the sea level was 30m lower. This river had to be feed by other rivers to obtain the necessary volume to exist. Therefore, the rivers that feed the Thames were also ten times larger than the same rivers today. This could only happen if the ground water levels were higher.

timeline

4500 BCE	2500 BCE	800 BCE	0-400 AD	1-2000 AD
Neolithic Age Begins	Bronze Age Begins	Iron Age Begins	Roman Period	Written History

Case study – RIVER OUSE

The Ouse is one of the four rivers that cut through the South Downs. It is presumed that its valley was cut during a glacial period, since it forms the remnant of a much larger river system that once flowed onto the floor of what is now the English Channel. The extent of the inundation it experienced after the last ice age can be observed in the lower valley which would have flooded; there are raised beaches 40 metres (Goodwood-Slindon) and 8 metres (Brighton-Norton) above present sea level. The offshore topography indicates that the current coastline was also the coastline before the final deglaciation, and therefore the mouth of the Ouse has long been at its present latitude, but as the height was up to 40m higher the width of the Ouse would have been like our other study the river Thames, ten times wider at its narrowest.

Archaeological evidence points to prehistoric dwellers in the area. Scholars think that the Roman settlement of Mutuantonis was here, as quantities of artefacts have been discovered in the area. The Saxons built a castle, having first constructed its motte as a defensive point over the river; they gave the town its name. But the most interesting aspect of Lewes is that it is seven miles from the current shoreline at Newhaven, until recent history (last 2000 years) it was the main port of the South coast as it had a huge natural shelter inland of the stormy seas.

Today the River Ouse is just a sleepy river running through the café bar area of the town yet in the recent past the town would have been a fishing port with boats moored up next to what we see as the High Street today.

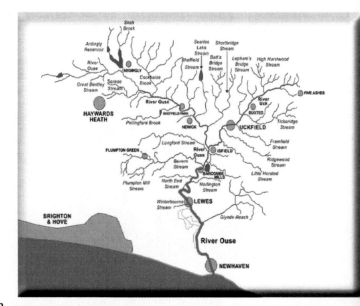

According to geologists the sea levels have been rising since the great flood after the last ice age and that the Isostatic uplift from the same event is now in reverse and lowering the landscape to compound the sea level rises - so why is Lewes gone from a sea port to a small idyllic river with café bars?

At Domesday (1086), the Ouse valley was still a tidal inlet with a string of settlements located at its margins. In later centuries the Ouse was draining the valley sufficiently well for some of the marshland to be reclaimed as highly prized meadow land. At this point in history the outlet of the Ouse at Seaford provided a natural harbour behind the shingle bar.

However, by the 14th century the Ouse valley was still regularly flooding in winter, and frequently the waters remained on the lower meadows through the summer. In

17000 BCE   You are HERE   **10000** BCE

Ice Age Ends   Mesolithic Age Begins

River Port of Lewes in the Late Neolithic Period

1422 a Commission of Sewers was appointed to restore the banks and drainage between Fletching and the coast, which may indicate that the Ouse was affected by the same storm that devastated the Netherlands in the St Elizabeth's flood of 1421. Drainage became so bad that 400 acres (1.6 km2) of the Archbishop of Canterbury's meadow at Southerham were converted into a permanent fishery (the Brodewater) in the mid-15th century, and by the 1530s the entire Lewes and Laughton Levels, 6,000 acres (24 km2), were reduced to marshland again.

Prior Crowham of Lewes Priory sailed to Flanders and returned with two drainage experts. In 1537 a water-rate was levied on all lands on the Levels to fund the cutting of a channel through the shingle bar at the mouth of the Ouse (below Castle Hill at Meeching) to allow the river to drain the Levels. This canalisation created access to a sheltered harbour, Newhaven, which succeeded Seaford as the port at the mouth of the Ouse.

The new channel drained the Levels and much of the valley floor was reclaimed for pasture. However, shingle continued to accumulate and so the mouth of the Ouse began to migrate eastwards again. In 1648 the Ouse was reported to be unfit either to drain the levels or for navigation. By the 18th century the valley was regularly inundated in winter and often flooded in summer.

So we can see quite clearly, even with rising sea levels the natural groundwater under this part of the South Downs is 15,000 years after the great melt and flooding still causing problems and it's only the industrial drainage of this area still keeping it from flooding. But was the flooding from the sea or from freshwater from groundwater under the South Downs.

This is an extract from a report on the viability of a new desalination plant in Newhaven. From 'Geoarchaeological Assessment' - (partial transcription from M. Bates in Dunkin 1998)

The Ouse basin is a medium sized catchment draining southwards in the southern Weald. The town of Newhaven and the investigation area lies at the mouth of the river where the lower course of the river has eroded a channel through the Chalk of the South Downs. The Pleistocene history of the river is poorly understood (Burrin and Jones, 1991) and remnants of older Pleistocene sediments (fluvial) occur upstream

timeline

4500 BCE	2500 BCE	800 BCE	0-400 AD	1-2000 AD
Neolithic Age Begins	Bronze Age Begins	Iron Age Begins	Roman Period	Written History

in the vicinity of the Brooks. Within the area of study Pleistocene deposits are restricted to head and valley fill sediments found on the valley sides and within the dry valley systems.

Holocene sediments dominate the valley bottom from the Brooks southwards to the sea at Newhaven. These deposits (mapped as alluvium by the British Geological Survey) include a wide variety of sediment types including marine beach gravels and sands, clay-silts, organic silts, peats and fluvial gravels and this zone has been described as the perimarine area (Burrin and Jones 1991).

Previous works indicated that some of these deposits contained freshwater and estuarine shells and that others reported pollen from these sediments (Thorley 1981) and a radiocarbon date of 6290+/-180 B.P., from a depth of –8.15m O.D. was reported from the Brooks (Jones 1971).

So geologists admit that the history of the Ouse (and other post glacial rivers) are 'poorly understood' which is the closest to a geologist admitting that the evidence does not fit the current theory. The evidence clearly shows that in 4290 BCE the Ouse valley was covered in freshwater. The only place to get this volume of freshwater would be the groundwater under the South Downs which must have still been so high that the valley remained flooded for over 10,000 years and not the few hundred year's geologists currently suggest.

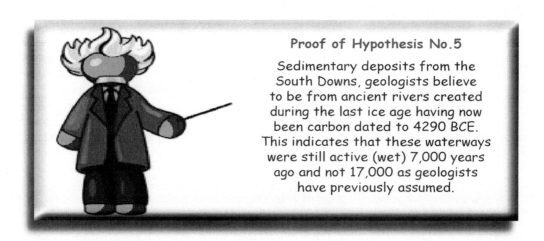

Proof of Hypothesis No.5

Sedimentary deposits from the South Downs, geologists believe to be from ancient rivers created during the last ice age having now been carbon dated to 4290 BCE. This indicates that these waterways were still active (wet) 7,000 years ago and not 17,000 as geologists have previously assumed.

17000 *BCE*   You are HERE   10000 *BCE*

Ice Age Ends   Mesolithic Age Begins

Chapter 4 - Geological Maps of Britain

Superficial deposits are overlain on bedrock geology ("solid geology" in the terminology of maps) is a varied distribution of unconsolidated material of more recent origin. It includes material deposited by glaciers (boulder clay, and other forms of glacial drift such as sand and gravel). "Drift" geology is often more important than "solid" geology when considering building works, drainage, sitting water boreholes, sand and gravel resources and soil fertility. Although "drift" strictly refers to glacial and fluvial-glacial deposits, the term on geological maps has traditionally included other material including alluvium, river terraces, etc. Recent maps use the terms "bedrock" and "superficial" in place of "solid" and "drift" 4

Clear as mud really!!

Archaeologists have to rely on radiocarbon dating evidence to date prehistoric sites – no other credible method currently exists. Geologists however, do not use such methods and rely on dating by 'strata' evidence alone, as the rationale is that new soil is built upon the old strata (like rings on a tree) allowing geologists to date the strata.

The problem with this method is that if the surface in question is a river, the layers are interrupted by the ebb and flow of the water thus confusing matters. A recent study into perceived strata date did not match the real carbon dating evidence as shown in our next case study on the Mississippi delta.

Case Study - Anomalous Radiocarbon Dates

Problems associated with radiocarbon dating which can be both too old and in the wrong sequence, can be found in Annual Review of Earth and Planetary Sciences (Stanley,2001).

"ANOMALOUS RADIOCARBON DATES: In contrast with patterns of clustered radiocarbon dates at the base of Holocene sections, there is a weaker relationship between C-14 dates and core depths throughout most deltaic core sections. This poor relationship has been observed since early applications of the radiocarbon dating method to Mississippi Delta cores (Fisk & McFarlan 1955, Frazier 1967). A review of the literature indicates that most deltas for which radiocarbon dates are available, regardless of geographical and geological setting, record this inconsistent up section as stratigraphy.

Radiocarbon dates, both conventional and accelerator mass spectrometric (AMS), are not— as expected — consistently younger up core between the base and surface of deltaic sequences. In addition to age-date reversals up core, some dates in Holocene sections are clearly too old (some too late Pleistocene in age) and, not infrequently, those near the upper core surfaces are of mid- to late Holocene age.

In general, there is a modest to poor—and in some cases no—relationship among C-14 dates, core surface elevation, subsurface depth of sample in the Holocene sequence, material used for dating (i.e. shell, organic-rich sediment, and peat), and geographic position of a core site relative to the delta coast."

Clear as mud really!!

timeline

4500 BCE	2500 BCE	800 BCE	0-400 AD	1-2000 AD
Neolithic Age Begins	Bronze Age Begins	Iron Age Begins	Roman Period	Written History

As reported by Delibrias in 1989, "These findings are both remarkable and disturbing, because they call into question the reliability of both dates and method; they raise a concern regarding use of the radiocarbon method as presently applied to deltas. A literature survey indicates that deltas are by no means the only late Pleistocene to Holocene settings were dating problems are encountered.

Consequently, based on the fact that, as detailed above, geologists they question their own methodologies, it would be sensible to question the dates estimated by archaeologists using this same method of radiocarbon dating as with site on or near water. Evidence must be 'put under the microscope and re-examined' rather than simply assumed to be correct. Let's look at our main case study site at Stonehenge and review the way geologists date the 'dry river valleys' of this area.

Not confusing at all really!

If we look at the geological map of this area, we notice that it is covered with what looks like rivers of 'pink' and 'yellow'. The pink shows what is termed as 'head' and is defined as 'variable deposits of sand, silty clay, local gravel, chalky and flinty in dry valleys'. The yellow is defined as 'Alluvium, clay, silt and sand, locally organic with gravel'.

Not a lot of difference really except the alluvium and what is that you may be asking? Well it is defined as 'Alluvium is typically made up of a variety of materials, including fine particles of silt and clay and larger particles of sand and gravel'.

Not confusing at all really!

So when sand, clay and gravel is not part of a dead dried up water system hundreds of thousands of years old it's called 'head' or 'hill wash' but not 'alluvium', which is the remains of an ancient river.

This cannot be proven and is clearly utter nonsense as its impossible to differentiate between the two.

The reality is that they are both part of the same thing, an ancient water system of an age as yet unknown. What is more important is that these maps prove that once upon a time there was water in the dry valleys that surrounded Stonehenge. The next question we need to consider is how large were these waterways as the

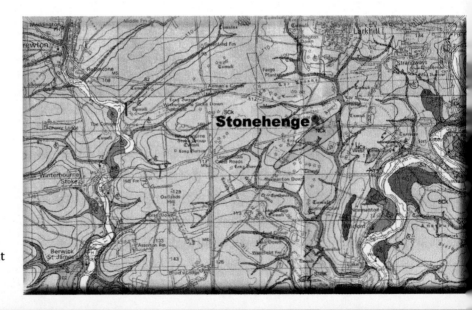

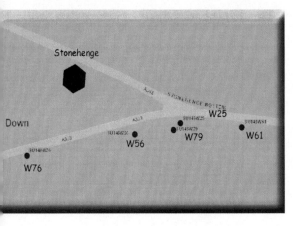

maps, show how wide spread these waterways were stretching from the coastline all the way inland to places like Stonehenge and Avebury. But how large were they as the maps show them to be quite narrow?

From the British Geological Society (BGS) website we have obtained data of boreholes in order to judge the extent of the river width at the site's base. According to maps provided by the BGS this ancient waterway has a thin (80 meters) 'head' at the Stonehenge bottom of our main case study site. But the boreholes tell us something completely different. We have taken the readings from boreholes made for the old proposed diversion of the A303 and examined them in detail to find out the real extent of these deposits in the dry river valleys and the results are quite astonishing.

According to the geologists, the river wash or head affected just eight boreholes in the Stonehenge bottom region of the A303 (SU14SW25, SU14SW79, SU14SW80, SU14SW99, SU14SW97, SU14SW60) and indeed looking at SU14SW25 in detail the diagram shows clearly the 'topsoil' again made up of the same sand, clay and gravel is 0.80m deep and the head is another 0.4 m deep, yet another borehole just 36m away shows just 0.22 of topsoil and just 0.17m of head, four times smaller. If we look at boreholes outside the map region such as SU14SW61 - 210m to the east of the centre of the maps main river wash deposit we find NO TOPSOIL or HEAD, in the opposite direction SU14SW56 240m to the west we find the same.

This would suggest the map is correct and the head deposit maybe as suggested hill wash - but let's take a closer look!

The deposits found at the two extremes have what is known as Chalk Grade V deposits - so what is that you ask yourself? Well the description is ' structureless chalk composed of slightly sandy silty subtriangular fine to course gravel, weak to medium in density, white with occasional black speckling, matrix is light brown'.

To put this into layman's terms, you have 'structureless chalk' - it is broken and has been moved from the chalk bed, probably by a river cutting into it. Sand silt and gravel we now know is head or alluvium, whatever you wish to call it and the matrix is brown in colour which you would expect from the residue of a river bank. The bedrock is Chalk and chalk is solid and white - so how far down do we need to go to find the bedrock untouched by water?

On the borehole to the east SU14SW61 silt and sand is found in the chalk all the way down to 6m, even at 12m down the report reads of "occasionally thinly in filled with silty sandy comminuted (broken) chalk".

timeline

4500 BCE	2500 BCE	800 BCE	0-400 AD	1-2000 AD
Neolithic Age Begins	Bronze Age Begins	Iron Age Begins	Roman Period	Written History

Are they really suggesting that the ancient river did not flow near over this area?

And what of the other borehole SU14SW56? At 14m below the surface we still do not find the bedrock but "weak to very weak thinly bedded light brown to brown silty fine to medium grained sandstone. Silt of chalk, sand is sub-rounded... in filled with brown silty sand". Again this shows that groundwater was on this spot in prehistoric times. In fact if we go to the extremes of the borehole area SU14SW76 some 500m to the west of the maps centre we even find the return of topsoil, head and structureless chalk, down to 1.1m.

Clearly the prehistoric river was not just 80m wide but more like 1000m wide, better known as a kilometre. So armed with this proof of evidence we can move forward to show how these rivers affected our prehistoric past, putting aside the idea of 'hill wash' as being more 'eye wash' than real substance.

Moreover, geologists had noticed that from a topographical perspective, 'River Terrace Deposits' for the River Avon had, in certain areas, the highest terraces in the district of Salisbury, which were between 30m and 45m above the floodplain of the lower Avon – in other words, at some point in the past, the River Avon flowed approximately 40 metres higher than it does today, the height of a small office block. One of these River Terraces has been carbon dated some 300m from Durrington Walls, Scaife (in Cleal and Pollard 2004) reported pollen found in peat and alluvium sediments had a carbon date of 8300 - 7200 BCE (GU-3239) support our observations of this area. If this level of flooding was reproduced today, it would flood Stonehenge Bottom and make Stonehenge a peninsula.

Borehole Data from Stonehenge Bottom

Peat

Peat (turf) is an accumulation of partially decayed vegetation. One of the most common components is Sphagnum moss, although many other plants can contribute. Soils that contain mostly peat are known as a histosol. Peat forms in wetland conditions, where flooding obstructs flows of oxygen from the atmosphere, reducing rates of decomposition.

Bogs are the most important source of peat, but other less common wetland types also deposit peat, including fens, pocosins, and peat swamp forests. There are many other good words for lands dominated by peat including moors, muskeg, or mires. Landscapes covered in peat also have specific kinds of plants, particularly Sphagnum moss, Ericaceous shrubs, and sedges (see bog for more information on this aspect of peat). Since organic matter accumulates over thousands of years, peat deposits also provide records of past vegetation and climates stored in plant remains, particularly pollen. Hence they allow humans to reconstruct past environments and changes in human land use[5]

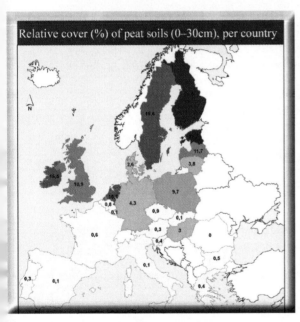

Relative cover (%) of peat soils (0–30cm), per country

Peat forms when plant material, usually in marshy areas, is inhibited from decaying fully by acidic and anaerobic (lack of oxygen) conditions. It is composed mainly of marshland vegetation: trees, grasses, fungi, as well as other types of organic remains, such as insects, and animal remains. Under certain conditions, the decomposition of the latter is inhibited, and archaeologists often take advantage of this.

Peat layer growth and the degree of decomposition (or humidification) depend principally on its composition and on the degree of waterlogging. Peat formed in very wet conditions accumulates considerably faster, and is less decomposed, then that in drier places. This allows climatologists to use peat as an indicator of climatic change.

Unlike sub-soils, such as head and alluvium, most peat bogs can be accurately carbon dated - this is central to my hypotheses. The result of carbon dating concludes that peat bogs in Britain were formed after the flooding caused by glaciers melting at the end of the last Ice Age over 10,000 years ago.

timeline

4500 BCE	2500 BCE	800 BCE	0-400 AD	1-2000 AD
Neolithic Age Begins	Bronze Age Begins	Iron Age Begins	Roman Period	Written History

If my hypothesis is correct, there should be more peat here in Britain than in similar countries to the South, as the Isostatic Transformation (landscape compression contributing to the flooding) would have a higher percentage of peat as these islands (that we now know as Britain and others to the East and North East of Europe) were affected to a greater degree by raising Groundwater tables and making them marshier, wet and waterlogged more than other less affected countries.

According to the European Soil Bureau Network 10.9% of the total land surface in Britain is peat. In comparison, France, our nearest Southerly neighbour, has only 0.6% and Spain, a mere 0.1% although they share very similar sub-soils and prehistoric foliage. This clearly shows that Britain has 18 to 100 times greater peat deposits than any of its Southerly neighbours.

The reason for this inconsistency is that the ice cap lay on the North West side of Britain during the Ice Age. As the ice melted, Britain was consumed by the flood waters which didn't reach past the English Channel and within Britain itself, the North West region suffered from the greatest amount of flooding. Consequently, the greatest deposits of peat can be found in the North West region of Britain.

In addition, the ratio of peat to land mass in Ireland is an incredible 16.5% - almost the same as that found in Scotland, 16.3%. Remember, both counties would have taken the full weight of the ice cap during the last Ice Age, together with the North West region of Britain.

Proof of Hypothesis No.6

Abnormally high peat deposits in Britain, compared to our 'southerly' neighbouring countries bears testament that the British landscape was in recent times (last 9,000 years) must have been flooded for an extensive period.

Case Study - Star Carr

Star Carr now lies under farmland at the eastern end of the Vale of Pickering. During the Mesolithic the site was near the outflow at the western end of a paleolake, known as Lake Flixton. At the end of the last ice age a combination of glacial and post-glacial geomorphology caused the area drained to the west (away from the shortest-distance to the sea at Filey). The basin filled by Lake Flixton was probably created by glacial 'scarring'.

The site is preserved due to Lake Flixton in-filling with peat during the course of the Mesolithic. Waterlogged peat prevents organic finds from oxidising and has led to some of the best preservation conditions possible (such conditions have preserved the famous bog bodies found in other parts of northern Europe). As a result of such good conditions archaeologists were able to recover bone, antler and wood in addition to the flints that are normally all that is left on sites from this period. The finds in the peat are dated to 8700 BCE[6]

In conclusion

We have shown that if we delve into the deep geology of Britain we find that it has suffered a catastrophic disaster during and after the last ice age which we are just recovering from. This climatic disaster left the land we see that is dry and green today as a land of islands that was surrounded with lakes and gigantic rivers for thousands of years after man had returned to these shores.

This process can be seen in our last case study on the Somerset Flats as it not only shows that the raised groundwater levels after the last ice age continued for some considerable time and that archaeological artefacts found in this period are causing major problems in understanding as the traditional models do not reflect the reality of our history.

Case Study - Somerset Flats

The Somerset flats are a strange paradox to any rational person studying the area. A paper by the Somerset County Council on the Palaeolithic and Mesolithic Period clearly states that;

The Early Mesolithic (as defined for present purposes) covers most of the first "epoch". In the Late Glacial and Early Holocene, the ameliorating Palaeolithic and Mesolithic climate was reflected by a rapid rise in sea level of 1cm per year, with a drop in this rate after c.7000–6500 BP (c.5990–5350 cal BCE).

Sea level rose from 35m below present mean sea level (MSL) at c.9500 BP (c.9130–8630 cal BCE), reaching 5m below MSL in the Bristol Channel by c.4000–3800 cal BCE (the rate of rise having slowed by c.4000 cal BCE).

At the time of the Mesolithic, the Somerset Plain was between 5m and 35m above the sea level - so why was it flooded?

Let's see what Wikipedia makes of this paradox?

Let's see what Wiki makes of this paradox?

The Somerset Levels, or the Somerset Levels and Moors as they are less commonly but more correctly known, is a sparsely populated coastal plain and wetland area of central Somerset, South West England, between the Quantock and Mendip Hills. The Levels occupy an area of about 160,000 acres (650 km2), corresponding broadly to the administrative district of Sedgemoor but also including the south-eastern part of the Mendip district. The Somerset Levels are bisected by the Polden Hills; the areas to the south are drained by the River Parrett, and the areas to the North by the rivers Axe and Brue. The Mendip Hills separate the Somerset Levels from the North Somerset Levels. The Somerset Levels consist of marine clay "levels" along the coast, and inland (often peat-based) "moors"; agriculturally, about 70 percent is used as grassland and the rest is arable. Willow and teazel are grown commercially and peat is extracted.

				timeline
4500 *BCE*	**2500** *BCE*	**800** *BCE*	**0-400** *AD*	**1-2000** *AD*
Neolithic Age Begins	Bronze Age Begins	Iron Age Begins	Roman Period	Written History

One explanation for the county of Somerset's name is that, in prehistory, because of winter flooding humans restricted their use of the Levels to the summer, leading to a derivation from Sumorsaete, meaning land of the summer people. A Palaeolithic flint tool found in West Sedgemoor is the earliest indication of human presence in the area. The Neolithic people exploited the reed swamps for their natural resources and started to construct wooden trackways, including the world's oldest known timber trackway, the Sweet Track, dating from the 3800 BCE. The Levels were the location of the Glastonbury Lake Village as well as two Lake villages at Meare Lake. Several settlements and hill forts were built on the natural "islands" of slightly raised land, including Brent Knoll and Glastonbury. In the Roman period sea salt was extracted and a string of settlements were set up along the Polden Hills[7]

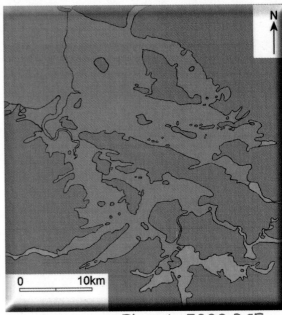

Somerset Flats in 5000 BCE

Winter Flooding - so how do we know that?

The most recent paper by Dr Richard Brunning shows that in 8300 BCE the entire region was covered with over 100 islands. But if the Sea level was below the land surface - by at least 5m, how did it flood, even in winter? The amount of rain needed to flood this area is equivalent to a far eastern monsoon - and if it rained that much in Somerset, it must have rained the same throughout Britain?

The only logical answer is that the rivers feeding this 'delta' were flooded with excess water from the ice age that had raised their groundwater levels by about 30m. This would allow sufficient groundwater to flood the Somerset Levels before flowing into the Sea some 5m below the land level[8].

What is more interesting is that the same site showed us quite clearly why academic argument needs to be questioned and should never be taken as correct.

Somerset was the site of the UK's oldest open-air cemetery, the county council says;

Recent radiocarbon dating of two skulls found at a sand quarry in Greylake nature reserve near Middlezoy in 1928 revealed them to be 10,000 years old. The council said the find was made under its Lost Islands of Somerset project by a team investigating the archaeology of the Somerset Levels. Since their discovery, the skulls have been held at Bridgwater's Blake Museum.

The new findings show that by around 8,300 BCE, hunter-gatherers were burying their dead on what was once an island amid the Levels. All the other human remains from this early period in Britain have been found in caves such as Aveline's Hole in Somerset which is the largest Mesolithic burial ground in the UK.

Somerset County Councillor Christine Lawrence, cabinet member for community services said: "Somerset's

17000 *BCE*   **10000** *BCE*

Ice Age Ends   Mesolithic Age Begins

Proof of Hypothesis No. 7

The Somerset flats are well known to have flooded in the recent past because of sea intrusion. But during the Mesolithic Period, the sea level was 5-35m lower than today. The 100+ islands reported by Dr Brunning can only exist if the water was fresh water from local rivers, due to higher ground water levels.

wonderfully rich heritage plays a big part in attracting visitors. I'm delighted that this project has thrown new light on to these exciting finds."

"This was amazing news and was just the result we were hoping for," added Dr Richard Brunning, from Somerset County Council's Heritage Service who is leading the Lost Islands of Somerset Project.

"It shows that a Mesolithic hunter-gatherer group was operating from the island and burying its dead there. Such open-air cemeteries are extremely rare in Europe and this is the only one known from the UK."

Flint tools were also found in large numbers on the site in the 1950s suggesting that it was used as a long-term camp site. The Lost Islands of Somerset Project team will carry out more analysis on the skulls and tools to ascertain how this ancient community lived and died[9]

Quite interesting until you read between the lines and then some disturbing questions arise; these skulls were discovered in 1922 - **why is it taken 80 years to realise that the find was important?** The good doctor claims that they were a burial - so did they find the rest of the skeleton and if so what position where they buried in? What ages were the skulls - are we looking at a family or ancestors over a period of time?

So what can we truly assume from this evidence?

Well looking at the online archive I can tell you that only the skulls were found so burial is unlikely and as the original discoverer believed them to be victims of the Sedgmoor battle burials of the 16th century - we can now see why it took 80 years for someone to carbon date the skulls.

As you see when archaeologists get it wrong - they can do it big time!

The doctor talks of the skulls being buried on Somerset islands. Unfortunately I have found the exact site where the skulls were found and the profile shows that the quarry is not on an island - nice try!! What they have found is a result of a boat accident in 8300 BCE, as you can see these would be islands close to the

timeline

4500 BCE	You are HERE	2500 BCE	800 BCE	0-400 AD	1-2000 AD
Neolithic Age Begins		Bronze Age Begins	Iron Age Begins	Roman Period	Written History

site and no doubt they were either sailing to or from Glastonbury via Priest Hill which contains a Long Barrow at Collard Hill.

Sorry to say, just another case of when academics just get it wrong due to them trying to tow the accepted path of scientific nonsense rather than looking at the facts and coming to their own conclusions. This is not the first time that the 'experts' have got it all wrong - the Visitors Car Park Post Holes found in 1966 and classified as late Neolithic (the same date as Stonehenge) - it was an eagle eyed student (without a PhD or MSc) who realised that the wood found, pine did not exist in the late Neolithic Period - it was only then that the samples were sent for carbon testing. Where it would still be sitting now and my hypothesis would have no real evidence.

How many other samples have yet to be discovered?

This is why I question all academic research, not because they don't understand but they want to keep their jobs and be employed in the future as they have mortgages to pay and children to feed, so they will reflect the 'accepted dates and theories'. This is institutional self regulation - we see it with the financial services and with the media. It is neither the truth nor full disclosure.

Why has it taken 80 years to realise that the find was important?

17000 BCE   You are HERE   10000 BCE

Ice Age Ends   Mesolithic Age Begins

Section Two: ARCHAEOLOGICAL SECTION

The following section looks at the existing archaeological evidence within prehistoric sites in Britain to see if there is further evidence to prove my hypothesis. Surprisingly, the most undeniable evidence comes from past archaeological excavations, the findings of which have been misinterpreted. These misinterpretations are due to archaeologists' fixation with convention so when an 'unusual' find appears, it is not fully recognised as it does not fit into accepted archaeological conventions.

Now we have established that there is compelling evidence of prehistoric groundwater, as shown on the geological maps of Britain, we must now look to see if we can find more realistic dates for these waterways. The most effective way in establishing if the raised groundwater levels affected our ancestors is to look at the landscape features to see if they have evidence of this groundwater in there construction.

We have concentrated on the Stonehenge region as it is the only area in Britain that provides sufficient detail to be able to test my hypothesis. Nevertheless, I have also found similar evidence throughout Britain, perhaps less detailed but nonetheless firm evidence, which will be offered in further books, blogs and articles.

Wansdyke

timeline

4500 BCE	2500 BCE	800 BCE	0-400 AD	1-2000 AD
Neolithic Age Begins	Bronze Age Begins	Iron Age Begins	Roman Period	Written History

Chapter 5 - Dykes and other Earthworks

The modern word dike or dyke most likely derives from the Dutch word "dijk", with the construction of dikes in the Netherlands well attested as early as the 12th century. The 126 kilometres (78 mi) long Westfriese Omringdijk was completed by 1250, and was formed by connecting existing older dikes. The Roman chronicler Tacitus even mentions that the rebellious Batavi pierced dikes to flood their land and to protect their retreat (AD 70). The word dijk originally indicated both the trench and the bank[10.]

If you study archaeology at university or even on a ordinance survey map at length, you will notice strange earthworks on the sides of hills of Britain, with no rational explanation as to why there are where they are. At university, these features are mostly ignored or an excuse is made for their construction. The reality is that these features do not make any sense unless there is another factor in operation we have ignored or are oblivious to.

The first thing to notice is that the word 'Dyke' is associated with water. It does seem strange you would call an earthwork on top of a hill a Dyke, unless there was some history passed down through the years to its real use. If we look at the most famous Dyke in Britain 'Offa', we notice that it is attributed to a Saxon King and therefore could not be prehistoric. Or is this a clear indication of how archaeologists find excuses for these features rather than true empirical evidence?

"Offa's Dyke (Welsh: Clawdd Offa) is a massive linear earthwork, roughly followed by some of the current border between England and Wales. In places, it is up to 65 feet (19.8 m) wide (including its flanking ditch) and 8 feet (2.4 m) high. In the 8th century it formed some kind of delineation between the Anglian kingdom of Mercia and the Welsh kingdom of Powys."11

At face value this seems to answer all the questions about this dyke - except the water connection. But if you delve further down to look at the evidence such as findings from the dyke and any written history you get a different version, for the Roman historian Eutropius in his book, Historiae Romanae Breviarium, written around 369 (AD), mentions the Wall of Severus, a structure built by Septimius Severus who was Roman Emperor between 193 and 211 (AD):

So the Romans built it 700 before Offa.. or did they?

"He had his most recent war in Britain, and to fortify the conquered provinces with all security, he built a wall for 133 miles from sea to sea. He died at York, a reasonably old man, in the sixteenth year and third month of his reign."

So the Romans built it 700 years before Offa.... or did they?

17000 BCE	You are HERE	10000 BCE
Ice Age Ends		Mesolithic Age Begins

For they are now finding Neolithic flints inside the ditch of the dyke - so how did they get there? As we will show in our case study on Old Sarum in this section, the Romans are famous for taking existing features, such as ditches and adding a defensive bank for their own use as did the Normans who followed them some time later in history. So Offa's Dyke has nothing to do with Offa, but this archaeological reality or misinterpretation is the key to why our history is not as we perceive.

But that is not enough to prove the higher groundwater levels in prehistory contributed to these strange earthworks, so let's look at some in our study area where we have the prehistoric maps of the Mesolithic and Neolithic area.

CASE STUDY - Wansdyke

Wansdyke consists of two sections of 14 and 19 kilometres (9 and 12 miles) long with some gaps in between. East Wansdyke is an impressive linear earthwork, consisting of a ditch and bank running approximately east-west, between Savernake Forest and Morgan's Hill. West Wansdyke is also a linear earthwork, running from Monkton Combe south of Bath to Maes Knoll south of Bristol, but less impressive than its eastern counterpart. The middle section, 22 kilometres (14 miles) long, is sometimes referred to as 'Mid Wansdyke', but is formed by the remains of the London to Bath Roman road. It used to be thought that these sections were all part of one continuous undertaking, especially during the Middle Ages when the pagan name Wansdyke was applied to all three parts.

East Wansdyke in Wiltshire, on the south of the Marlborough Downs, has been less disturbed by later agriculture and building and remains more clearly traceable on the ground than the western part. Here the bank is up to 4 m (13 ft) high with a ditch up to 2.5 m (8.2 ft) deep. Wansdyke's origins are unclear, but archaeological data shows that the eastern part was probably built during the 5th or 6th century. That is after the withdrawal of the Romans and before the takeover by Anglo-Saxons. The ditch is on the north side, so presumably it was used by the British as a defence against West Saxons encroaching from the upper Thames Valley westward into what is now the West Country.

West Wansdyke, although the antiquarians like John Collinson considered West Wansdyke to stretch from south east of Bath to the west of Maes Knoll, a review in 1960 considered that there was no evidence of its existence to the west of Maes Knoll.   Keith Gardner refuted this with newly discovered documentary evidence. In 2007 a series of sections were dug across the earthwork which showed that it had existed where there are no longer visible surface remains. It was shown that the earthwork had a consistent design, with stone or timber revetment. There was little dating evidence but it was consistent with either a late Roman or post-Roman date. A paper in "The Last of the Britons" conference in 2007 suggests that the West Wansdyke continues from Maes Knoll to the hill forts above the Avon Gorge and controls the crossings of the river at Saltford and Bristol as well as at Bath.

As there is little archaeological evidence to date the western Wansdyke, it may have marked a division between British Celtic kingdoms or have been a boundary with the Saxons. The evidence for its western extension is earthworks along the north side of Dundry Hill, its mention in a charter and a road name.

timeline

4500 BCE	2500 BCE	800 BCE	0-400 AD	1-2000 AD
Neolithic Age Begins	Bronze Age Begins	Iron Age Begins	Roman Period	Written History

The area of the western Wansdyke became the border between the Romano-British Celts and the West Saxons following the Battle of Deorham in 577 AD. According to the Anglo-Saxon Chronicle, the 'Saxon' Cenwalh achieved a breakthrough against the British Celtic tribes, with victories at Bradford on Avon (in the Avon Gap in the Wansdyke) in 652 AD, and further south at the Battle of Peonnum (at Penselwood) in 658 AD, followed by an advance west through the Polden Hills to the River Parrett. It is however significant to note that the names of the early Wessex kings appear to have a Brythonic (British) rather than Germanic (Saxon) etymology.[12]

I thought I should give you all the 'considered' opinion of this dyke, just to show how confused and inaccurate our history books are today. Go get yourself a OS map and have a look at this Dyke.

But, before you do, remember Offa's Dyke, which carries the same Saxon name and estimated date?

Well again recent archaeologists have found in the bottom of one of the ditches evidence that can be carbon dated and the results - Radiocarbon laboratory of Queen's University, Belfast on 7 July 1997 (sample number UB-4158), is 1571 ± 69 years BP (Before Present).

It was there during the Roman period and predates the Saxon King Offa by about four hundred years. Moreover, Romans did not built pointless earthworks (as we will see from our OS map) therefore it must predate even the Roman Invasion.

So let's have a look at the map - to the East the ditch ends at the village of Cadley, which borders the forest of Savernlake. Now archaeologists would say 'perfect cover' for the end of your 'defensive ditch', except the forest was not there when the ditch was built as a Roman road goes through the centre and Romans did not put roads through the middle of forests because of ambush, so the forest was planted after the Roman invasion.

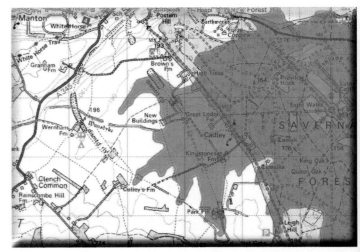

End of Wansdyke - East section

If you continued another 6km you would have reached water - a natural boundary, remembering you have already cut 19km, or turn south and that's just 3km. To the west it's even worse, you could go south again and connect to the river, using that as a natural defensive boundary, but it just stops. If you were to attack the ditch you'd be nuts as you can just walk around it as the German's did on the Siegfried line at the outbreak of world war two.

17000 BCE   You are HERE   **10000** BCE

Ice Age Ends   Mesolithic Age Begins

Now let us consider the manpower it needed to create such a structure. At 65 km in length (65,000m), its volume can be calculated as about 1.2 million m³ of soil and turf (if it is 2.5m deep and 7.5m wide as an average) at least, this is the approximate volume of material removed from the ditch. This is ten times greater than what was excavated at Avebury, which is estimated to take 1.5 million man -hours - so are we looking at 15 million man hours?

That would equate to 100 men working non-stop for 40 years to complete the task - they must have had a very good reason to this and clearly it was not for defence.

If we add the known prehistoric groundwater we have found then the absurd ditch becomes clear. For the ditch is in the middle of an island and perfectly cuts the island into two. This ditch is not a defensive earthwork it is what we would call a 'canal' that allowed boats to sail from one end of the island to the other in the Mesolithic period (8500 BCE - 4500 BCE). There are even signs as suggested by John Collinson (as above) that as the groundwater levels fell at the end of the Mesolithic Period about 4,500 BCE, they extended it to meet the Neolithic Shoreline.

But these canals are not limited to massive constructions like Wansdyke, they can be found within a stone's throw of Stonehenge and it's these smaller canals that give us our first hint to why they used them apart from the obvious transportation links.

Case study - Winterbourne Crossroads at Stonehenge

The Winterbourne Stoke Crossroads round barrow cemetery comprises a linear arrangement of 19 late Neolithic / Early Bronze Age circular earthwork monuments, commonly known as round barrows. Winterbourne Stoke 3 (Monument Number 870372) to 10 (870444) are aligned to the north-east of the Neolithic long barrow known as Winterbourne Stoke 1 (Monument Number 219696). They extend south-west / north-east for nearly 600m: this alignment continues after a gap of circa 100m (see Winterbourne Stoke 22: Monument Number 219720).

Proof of Hypothesis No. 8

Wansdyke has no military advantage as it seems to end without any defences on both edges. Yet when our proposed prehistoric waterways are introduced, both ends of this earthwork meets shorelines showing that it was in fact a canal.

timeline

4500 BCE	2500 BCE	800 BCE	0-400 AD	1-2000 AD
Neolithic Age Begins	Bronze Age Begins	Iron Age Begins	Roman Period	Written History

A roughly parallel secondary alignment immediately to the west comprises Winterbourne Stoke 2a (Monument Number 866648) to 12 (Monument Number 870446). A cluster of barrows sits slightly apart, circa 250m north-west of the main alignment (Monument Number 215072). Most of the barrows were excavated by Sir Richard Colt Hoare in the early 19th century[13.]

As you see from the above explanation from English Heritage the earthworks are not even mentioned let alone measured and excavated. This is because the most fundamental understanding of the environment at the time of these barrows construction is unknown or misunderstood. If they looked under the surface of the soil they would see that not too far away from this large number of barrows are the remains of a prehistoric river shoreline.

If you look at an OS map of this area you will see the modern crossroads that now lies in the middle of the barrow complex, this intersection is no accident of history. As we will explain in detail later in this section Round Barrows are markers for Neolithic pathways and therefore when these prehistoric roads cross, you will get a collection of barrows from separate directions. Morden roads are mainly built on the remains of former roads and former roads are built then again on their previous paths.

The closest earthworks to the winterbourne crossroads run North West (NW) to South East (SE) for about 500m, it then seems to meet another earthwork running NNW to SSE for about the same distance making the entire earthwork about one kilometre long. If the size is like our Wansdyke earthwork, we are looking at a 20,000 cubic m³ excavation taking about 20,000 hours using antler picks and stone axes, or 100 men taking about 20 days or one man taking 5 years (non-stop) - so no small farming feature as archaeologists would have you believe.

Sitting at a height of 110m above sea level it is a curious feature, it can't be defensive as you can walk around the edge and it can't be for animals as it stops nothing - fences are much cheaper and quicker to construct even in prehistoric times.

But if we place into the landscape the prehistoric groundwater levels something incredible happens.

With the Mesolithic groundwater level which would be at a height of around 95m - 100m the closer of the earthworks cuts into the shoreline of the river creating a canal. Even

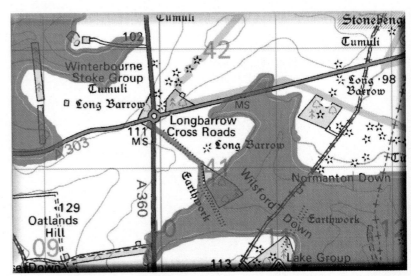

Longbarrow Cross Road showing 'Neo' Water levels

17000 BCE   You are HERE   10000 BCE

Ice Age Ends   Mesolithic Age Begins

more surprising is the second earthwork that was in a different direction cuts into the Neolithic waterway which is 10m lower at this point.

So why bother cutting a canal, where does it lead too?

Both canals have been cut to lead to one of the oldest monuments ever built, the Long Barrow. As we will discuss later the Long Barrow is where our prehistoric ancestors buried their dead and therefore the symbolic ritual of taking the body by boat to the Long Barrow for their final resting place is very important, in fact we still act out this same ceremony today with the slow funeral hearse drive to our cemeteries.

Clearly what we can see for the first time in history is the method and ideology of our ancestors bringing the body by boat to the shore where the Long Barrow was constructed. We can now estimate that initially (as not every Long Barrow has a canal associated to it) the shoreline was even higher at the time of the Long Barrow's construction (so we can estimate a more accurate date of construction) and when the groundwater receded the option to relocate was ignored for cutting a canal the 500m to the Long Barrow during the 4,000 years of the Mesolithic period - so it may have been done in small stages over many years.

Eventually, when the groundwater had dropped too low to keep the canal full, it was decided that it was more efficient to build a second Long Barrow on the shoreline. This again would have suffered from the drop of the groundwater level into the Neolithic, so a second canal was built from the shoreline of the Neolithic groundwater levels.

This one example of how the raised ground groundwater levels of prehistoric times not only can give us the reason for these strange landscape features, but also dates and sequences for these Long Barrows at winterbourne. Moreover, it gives us a clear understanding of the ceremonies and consequently the beliefs as they undertook rituals which have changed very little even today.

Proof of Hypothesis No. 9

The Winterbourne Crossroad Long Barrows are built originally on the ancient shorelines from our hypothesis. But as time went by the shoreline receded from these original points, so our ancestors dug 'canals' to join the Long Barrows to the receding shoreline.

timeline

4500 BCE	2500 BCE	800 BCE	0-400 AD	1-2000 AD
Neolithic Age Begins	Bronze Age Begins	Iron Age Begins	Roman Period	Written History

Chapter 6 – Ditches better known as Moats

Clearly, one of the most obvious pieces of evidence we can look at is within existing archaeological data. If groundwater was present in the past, some indication should still be present. Consequently, one of the more interesting facts found when studying any prehistoric monuments is that the constructors seemed to spend a disproportionate amount of time digging ditches to surround their monuments, whether around henges or barrows.

This practice would be considered strange, even if the prehistoric builders had effective and modern, labour saving tools. However, our ancestors only had the benefit of stone tools, bones and antler picks - making such excavation exceptionally slow, cumbersome and therefore even more bewildering.

Looking at Avebury (a henge monument), as an example, the most conservative archaeological estimation, suggests that the ditches surrounding Avebury, would have taken 1.5 million man hours to build. That's equivalent to 100 men working 12 hours a day, every day, for 3.5 years. In comparison, building a wooden palisade using the same tools, would have taken less than one month – merely 2% of the time and exhaustion.

Current archaeological theories surrounding these ditches maintain that they were used either as defensive fortifications and/or landscape features to keep out animals – not as a moat. This idea strikes me as somewhat flawed as a ditch (although in some parts over 5 metres deep) is significantly less effective than a palisade (a long line of sharpened wooden stakes planted into the ground), which would have been considerably easier, quicker to construct and more effective.

Lt-Col William Hawley was one of the amateur archaeologists, employed by the Ministry of Works to undertake excavations at our famous monuments Stonehenge and Avebury in the 1920s. Unfortunately, he was not the most 'careful' of archaeologists. This was a view shared by colleagues such as Atkinson. In his book, 'Stonehenge' 3rd edition, London 1979 - he suggests that Hawley's methods were somewhat 'inadequate'.

Despite this accusation of carelessness, he was still able to find some strange features, which can clearly be seen as evidence of a moat. For example, below a layer of chalk rubble infill (chalk which would have fallen naturally into the moat when it was disused) under a layer of flint, he discovered 'foot-trampled mud' – this, in an area of chalk land which has no natural mud/clay.

Now this in itself sounds quite interesting, if not conclusive as evidence of the existence of a moat, until you look for other landscape features with similar foundations which, when analysed, start to build up a much more conclusive picture. Such landscape features can be found in 'dew ponds'.

17000 BCE   **10000** BCE   You are HERE

Ice Age Ends   Mesolithic Age Begins

Neolithic Dew Pond

A dew pond is an artificial pond usually sited on the top of a hill, intended for watering livestock. Dew ponds are used in areas where a natural supply of surface water may not be readily available. The name dew pond (sometimes cloud pond or mist pond) is first found in the Journal of the Royal Agricultural Society in 1865. Despite the name, their primary source of water is believed to be rainfall rather than dew or mist.

The mystery of dew ponds has drawn the interest of many historians and scientists, but until recent times there has been little agreement on their early origins. It was widely believed that the technique for building dew ponds has been understood from the earliest times, as Kipling tells us in Puck of Pook's Hill. The two Chanctonbury Hill dew ponds were dated, from flint tools excavated nearby and similarity to other dated earthworks, to the Neolithic period.

They are usually shallow, saucer-shaped and lined with puddled clay, chalk or marl on an insulating straw layer over a bottom layer of chalk or lime. To deter earthworms from their natural tendency of burrowing upwards, which in a short while would make the clay lining porous, a layer of soot would be incorporated or lime mixed with the clay. The clay is usually covered with straw to prevent cracking by the sun and a final layer of chalk rubble or broken stone to protect the lining from the hoofs of sheep or cattle.

So was it a dew pond liner that Hawley found at the bottom of the ditches at Stonehenge?

I believe so. Nonetheless, Hawley can be forgiven for not recognising that these are clear features of a liner being used to stop lying water being absorbed through the porous chalk into the bedrock, because he was not looking for a moat. The surrounding area was perceived to have been dry for over half a million years and consequently Hawley could not have imagined that water would have existed.

Hawley also found evidence of what he called a strange 'dark soil layer' which existed within the primary fill, in many parts of the ditch. I would strongly suggest that this 'dark soil layer' consisted of decayed remains and sediment – further evidence of the existence of water. As time went by, organic matter would have floated to the bottom of the moats, especially if they had not been cleaned regularly. This would have left a lining of dark soil at the bottom of the moat when it fell into disuse and became dry.

4500 BCE	2500 BCE	800 BCE	0-400 AD	timeline 1-2000 AD
Neolithic Age Begins	Bronze Age Begins	Iron Age Begins	Roman Period	Written History

In 1923 Hawley reported that this 'ubiquitous dark layer' was found throughout his excavations, sometimes up to 8" thick, suggesting its existence for a considerable length of time, and it rested directly at the bottom of the ditch.

So, it's been established that Stonehenge had, at some stage, a layer of waterproofing, the 'layer of foot-trampled mud', added to the ditch. It's also been established that this ditch was, at some point, filled with water for quite some considerable amount of time, as evidenced by the deep sediment, 'dark layer', discovered.

But that's not all!

Proof of Hypothesis No. 10

Hawley found evidence of a 'water sealant' in the bottom of the ditch at Stonehenge, described as a 'layer of foot-trampled mud'. This 'water sealant' consisted of a mixture of clay/mud, chalk and struck flint. This arrangement is only usually found at the bottom of Mesolithic/Neolithic period 'dew ponds' – clear evidence of groundwater at Stonehenge, meaning that the ditch was a moat, not a defensive fortification.

If you look at remains of the Stonehenge moat today, it appears very shallow and uneven but when excavated it looks very different. It is, in fact, a series of 'individual pits', which, in places, are connected by shallow walls.

This opens up a series of fascinating unanswered questions:

- **Was the ditch half finished, for the pits differ in size and shape, or perhaps they we 'dodgy builders' who did not know what they were doing?**

- **If the ditch is indeed defensive as some archaeologists suggest for this and other henge sites, why not used the excavated chalk as a wall?**

- **Why leave internal walls with no purpose?**

If the ditch had been intended as a defensive feature, they would have used the chalk as a defensive bank, probably with a palisade on top.

But that's not all!

17000 BCE	10000 BCE	You are HERE
Ice Age Ends	Mesolithic Age Begins	

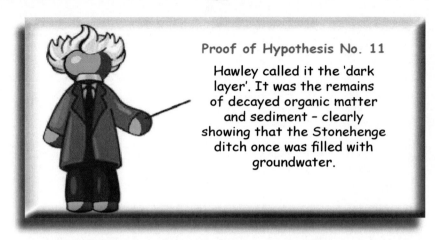

Proof of Hypothesis No. 11

Hawley called it the 'dark layer'. It was the remains of decayed organic matter and sediment – clearly showing that the Stonehenge ditch once was filled with groundwater.

The chalk excavated from the pits, however, was distributed evenly over both sides of the ditch – showing that defence was not their aim. A palisade was used at Stonehenge but not for defensive purposes as we will see later in this Section.

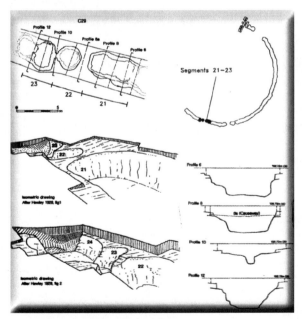

Stonehenge Ditch/Pits

So the ability to build defensive structures was available to our ancestors but on this occasion they decided not to use it as clearly it was built for another purpose and that purpose was as a groundwater filled moat. We can see now the only reason for cutting the ditch to different levels, which would be to search for the groundwater level under the soil. As chalk is a porous substance and water travels freely through it, the groundwater table over an area like Stonehenge can vary by a metre or so depending on the makeup of the chalk, as water runs through the fault lines in the chalk strata, leading to uneven groundwater levels over an area.

This variation of groundwater tables would also explain why the builders left shallow walls and why none of the walls go up to the surface – so the water could flow between the individual pits.

Bearing this in mind we can now answer one of the most frequently asked questions about Stonehenge – **why was it built on Salisbury Plain?**

timeline

4500 BCE	2500 BCE	800 BCE	0-400 AD	1-2000 AD
Neolithic Age Begins	Bronze Age Begins	Iron Age Begins	Roman Period	Written History

Proof of Hypothesis No. 12

Irregular ditches were deliberately cut at different depths and with low level internal walls to allow groundwater to flow over the ridges.

There are TWO answers to this question.

First, **CHALK!**

Our ancestors needed a place with a chalk sub-soil as it has two very special properties – it's porous and does not dissolve.

The problem with chalk is that it's difficult to excavate, in fact almost impossible. So if Mesolithic Man wanted an easy time and needed a large defensive ditch, he wouldn't have chosen soil, unless he's completely mad or had nothing better to do for the next ten years or so!

So clearly, our ancestors were attracted to chalk and its unique characteristics. Not only is it porous, but it is SLOW porous, which means that a pit or hole below the local groundwater table would gradually fill with or empty of water at a more consistent rate. In addition, because it's part of the limestone family it would not crumble and dissolve.

If these moats are used in conjunction with natural groundwater reeds (Phragmites australis is one of the main wetland plant species used for phytoremediation water treatment) the result is pure clean fresh water good enough to drink even in comparison to modern tap water. This type of water purification would not be seen in Britain for another 8,000 years.

The second reason is **WATER!**

We have seen from the previous Geological Section that the groundwater tables were far higher than we experience today, due to the ice caps compressing the landscape. For a site

Why was it built on Salisbury Plain?

17000 BCE

10000 BCE

You are HERE

Ice Age Ends

Mesolithic Age Begins

like Stonehenge this would mean that the monument would have been surrounded by groundwater on three sides.

This contributes to the groundwater being able to permeate into any pits or ditches dug below the groundwater table, and the water would be FILTERED! So the ditches that surrounded Stonehenge were deliberately cut by our ancestors below the groundwater table in order to allow them to fill with filtered, clean drinkable water.

It is commonly agreed amongst both geologists and archaeologist that 'dew ponds' could be used in areas without natural springs for drinking water so why didn't our ancestors simply build a "dew pond"– why go through the enormous and challenging task of building a 100 foot ring ditch full of freshwater pits?

The mystery of this ring ditch (which when filled with water becomes a moat) deepens when you look at the other ditches in our case study site Stonehenge, which shows that the entire site incorporates moats interlinking to each other.

Proof of Hypothesis No.13

The construction of a moat with a chalk sub-soil shows that the site was deliberately chosen to allow groundwater to freely flow into the moat at high tide – providing not only a groundwater filled feature but pure water, clean enough to drink.

The Heel Stone and two Station Stones.

The two Station Stone features found on the Stonehenge site are situated on the North West and South East side of the main monument within the ditched circle - both have individual moats surrounding them. This can be seen clearly in the Station Stone situated on the North West (also known as WA3595 and the North Mount), as discovered by 'Atkinson' in 1956. He observed that 'a small gully running East-West appears to lie beneath the rubble and earth bank'.

timeline

4500 BCE	2500 BCE	800 BCE	0-400 AD	1-2000 AD
Neolithic Age Begins	Bronze Age Begins	Iron Age Begins	Roman Period	Written History

This is a clear indication that the gully was a connecting strip, linking the main moat to the smaller moat surrounding the North West Station Stone. It wouldn't be an unrealistic suggestion to then say that the southern Station Stone was also connected to the main moat. If we are to believe traditional archaeologist's commonly held view that the moat surrounding the main monument at Stonehenge was constructed purely as a landscape feature, why would our ancestors connect the two smaller ditches to the main one?

The heel stone also has a ditch indicating its significance. This ditch is almost bisecting the avenue ditch, which we will show also was full of water. These ditches are invisible today as they were filled in long ago when the water left the site - but if they were important and ceremonial like the stones, why did they not keep the ditches from filling up with soil?

Proof of Hypothesis No. 14

These smaller moats that feed from the main moat at Stonehenge would only have been constructed if the main moat contained groundwater.

Stonehenge Moats

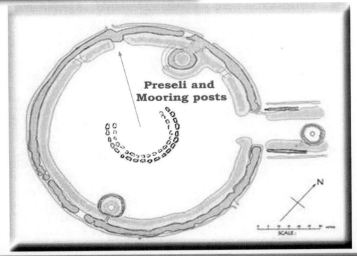

Preseli and Mooring posts

N

SCALE

Chapter 7 - Prehistoric Health Spa

Timothy Darvill, Professor of Archaeology at Bournemouth University, has recently revealed research that he believes shows that Stonehenge was an ancient healing place. This was further backed by subsequent episodes of Time Team. In his new book, 'Stonehenge: The Biography of a Landscape' the Professor cites that human remains excavated from burial mounds near Stonehenge, reveal that many of the buried had been ill prior to their death.

Darvill also suggests that these remains are not those of local people but of people who had come travelled from far and wide. For example, the Amesbury Archer, the name given to one of the remains identified, originated from what is now known as Switzerland. The Professor believes that Stonehenge would have been predominantly used during the winter solstice, when our ancestors believed it was occupied by Apollo, the Greek and Roman God of healing.

However, I would suggest that it was not the gods alone at Stonehenge that encouraged people from across the known world to travel such vast distances, it was another feature of Stonehenge that still survives today - Bluestones.

Bluestones are unexceptional, igneous rocks, such as Dolerite and Rhyolite. They are so called because they take on a bluish hue when WET. Over the centuries, legends have endowed these stones with mystical properties. The British poet Layamon, inspired by the folklore accounts of 12th Century cleric Geoffrey of Monmouth, wrote in 1215:

> The stones are great
>
> And magic power they have
>
> Men that are sick
>
> Fare to that stone
>
> And they wash that stone
>
> And with that water bathe away their sickness

As this ancient poem very obviously shows, the sick would BATHE away their illnesses. I find it surprising that neither Professor Darvill nor Time Team made this same discovery.

timeline

4500 BCE	2500 BCE	800 BCE	0-400 AD	1-2000 AD
Neolithic Age Begins	Bronze Age Begins	Iron Age Begins	Roman Period	Written History

Recently, findings by Professor Mike Parker Pearson of Sheffield University have revealed a smaller version of Stonehenge, confirming the link between Bluestones and WATER. The BBC reported that:

'About a mile away from Stonehenge, at the end of the 'Avenue' that connects it to the River Avon, archaeologists have discovered a smaller prehistoric site, named - appropriately, after the colour of the 27 Welsh stones it was made of – 'Bluehenge'. The newly discovered stone circle is thought to have been put up 5,000 years ago - which is around the same time work on Stonehenge began - and appears to be a miniature version of it.'

As the 'Bluehenge' site was constructed by the River Avon some 5,000 years after Stonehenge, it stands to reason that the original Stonehenge site, built for exactly the same purposes as 'Bluehenge', would also have been surrounded by water as well.

So how did our ancestors use the Bluestones for these healing treatments?

Another interesting aspect of the moat at Stonehenge discovered by Hawley was the number of craters found at the bottom of the moat. These craters were large enough to have accommodated quite large Bluestones. In fact, Hawley in one particular part of the moat found a two metre wide hole which he described as a post hole – this, however, could easily have been a stone hole as its size and shape was similar to the remaining standing Bluestones we see today in front of the Sarsen Stone circle.

It should be remembered that Bluestones aren't the same size as the Sarsen Stones, they're much smaller – in fact an average visitor to the Stonehenge Monument may quite easily scan over them without really noticing their presence. Archaeologists currently believe that there small size is because they are what remains after the damage by souvenir hunters over the years. But I propose that they may well have been small when brought to Stonehenge originally, for they have little to no building quality, but as a healing agent to be placed at the edge and in the moat, to initiate their healing qualities. As an indication of how these stones were originally used, archaeologists have identified a colossal amount of Bluestone chippings covering the entire site at Stonehenge, 3600 in fact.

Proof of Hypothesis No.15

The discovery of 'Bluehenge' by the River Avon proves that the Bluestones were located by water in order to be effective and provide their healing properties. Consequently, it is entirely feasible that Stonehenge, with its famous bluestone circle, must also have been constructed by to a water source.

17000 *BCE*

Ice Age Ends

10000 *BCE*

Mesolithic Age Begins

You are HERE

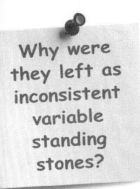

Why were they left as inconsistent variable standing stones?

I would suggest that just as we enjoy adding a variety of salts to our baths today, so did Mesolithic Man – he would have added a small amount of Bluestone chippings into the moat as he bathed. By chipping the Bluestones, revealing the inner core their healing qualities would have been enhanced.

The traditional view of why these quantities of Bluestones chippings are abundant, is because they were 'worked' upon and 're-shaped' to fit the chalk holes, into which they would be placed. This seems completely illogical - why would anyone in their right mind undertake the gruelling task of working on this exceptionally hard stone to fit into the holes, when it would have been so much easier to have dug the chalk soil first to accommodate the shape of the stones. Another traditional view is that the many chippings found were remnants from 'dressing' by re-shaping of these Bluestones.

If so, why were they left as inconsistent, variable standing stones – different sizes, different shapes?

We are aware that the larger Sarsen Stones were dressed on the inner side of the stone circle, as the flake marks are still visible - but no evidence to date that the bluestones also were dressed.

Given that archaeologists believe that the Bluestone chips exist only because of the re-working by our ancestors and they are the results of souvenir hunters, it would be interesting to compare their number (3600) to the number of chippings discovered from the softer, easier to break, more famous and more plentiful Sarsen Stones, which we know were re-worked.

You would think, proportionately, there would be approximately the same amount of chippings from each type of stone. Well, only 2170 Sarsen Stone pieces have been found despite there being over nine times more Sarsen Stones than Bluestones (251 cubic metres of Sarsen Stones v 28 metric metres of Bluestones).

So, assuming that there would have been a similar level of interest in Blue and Sarsen Stones by souvenir hunters, and a similar amount of re-working of the stones by our ancestors, you would expect to find at least 30,000 Sarsen Stone fragments, but as already shown a paltry 2170 have been discovered. Or if 2170 Sarsen stones pieces was the norm for both re-working and souvenir hunters - there should be only have been 240 bluestone fragments found not the 3600.

I can therefore very confidently conclude that the Bluestones were deliberately broken up to be used in the moat. Our ancestors are likely to have believed that once the outside covering of the Bluestones had been thoroughly exhausted, the beneficial properties would be diminished and so they were abandoned.

The concept of prehistoric man bathing away his ills may seem farcical to some, but throughout history it has been shown that mankind has been attracted to this type of treatment. It became

timeline

4500 BCE	2500 BCE	800 BCE	0-400 AD	1-2000 AD
Neolithic Age Begins	Bronze Age Begins	Iron Age Begins	Roman Period	Written History

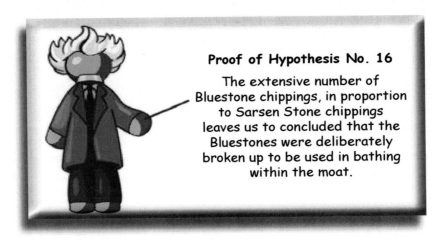

Proof of Hypothesis No. 16

The extensive number of Bluestone chippings, in proportion to Sarsen Stone chippings leaves us to concluded that the Bluestones were deliberately broken up to be used in bathing within the moat.

commonplace in Britain during the Roman Empire some 2,000 years ago when every large villa had its own spa. So is it really a giant leap to imagine that the origin of such activities could have been introduced at an early period?

When in the Bronze Age the moat at Stonehenge had eventually dried up and could no longer be used as a spa, these smaller Bluestones were abandoned and became scattered throughout the site. The larger Bluestones at the bottom of the dried moat were probably removed to the stone circle, explaining the huge variation in shapes and sizes of stone that we see at the Stonehenge monument today.

But what could be cured with this process?

Many of the diseases and illness of today did not exist in prehistoric times, the weak and deformed would probably die at childbirth and the mortality rate of children was probably high. So only the strong would normally live to an age where bathing was necessary - but what would be the common ailments that this process could cure?

The commonest form of death in prehistoric times would be infections from either injuries or simple cuts that are infected. We have cupboards full of antiseptics, but they did not. We will see later in the book this civilisation attempted with success surgery including amputation of limbs. The only way to survive such procedures will be to use antiseptic, and the best is rock salt.

As strange as this sounds, salt water is a highly effective antibiotic. Apart from organisms evolved to live in salty water, it is highly lethal to a large variety of common microbes. Next time you need to treat a sore throat, try gargling twice a day with a solution made from a teaspoon of common table salt dissolved in 300ml of lukewarm water. You will be amazed at the efficacy of this basic treatment. It goes without saying that salt water can be used as a topical treatment for other bacterial infections - just apply to the affected part for a few minutes at least twice a day[14]

Chapter 8 – Bluestones from far away

The most compelling evidence of the rise in water tables during the prehistoric period can be found in the car park of Stonehenge. Ignored by visitors who casually park their cars in the car park, three giant circles - similar to mini roundabouts - are painted on the floor. These painted circles show where post holes were discovered when the car park was constructed, and each measure approximately 1 metre in diameter. Interestingly, and rather ironically, the reason the car park was constructed at its current location as that archaeologists believed that area had no historical relevance –which is far from the truth.

Traditional archaeology describes the posts that would have been placed in these post holes as 'totem poles'. If this were true, however, why would Mesolithic Man have struggled with using huge trees, approximately 1 metre wide, for the simple purpose of erecting ritual 'totem poles'. As our ancestors only possessed flint axes and fire, it would seem more plausible that small trees would have been utilised to create these 'totem poles', rather than the giant metre wide variety as totem poles were used in North America to mark tribal territories.

If these were markers or religious symbols, would not the builders place them in a position of prominence on top of the valley?

The car park is 10m lower than the Stonehenge site and recent radiocarbon dating by Darvill and Wainwright's excavations in 2008, have confirmed from charcoal remains that the area on which the Stonehenge monument is situated was in use by 7200 BCE and this was confirmed a couple of years later by a team from the Open University who found an OX tooth eaten at a 'feast' that was dated also during the 7th millennium BCE at 6250 BCE, emphasising that is area of Stonehenge was used for other purposes, rather than an isolated site with marker posts.

Proof of Hypothesis No. 17

The post holes in the Stonehenge car park are approximately 1m in diameter and dug in a line, would reflect a shoreline, if the water table was higher. The only reason to construct post holes is to bear weight from above to allow the mass to dissipate within the hole. This proves that the holes were used as mooring stations with a cross piece being used as an ancient crane.

timeline

4500 BCE	2500 BCE	800 BCE	0-400 AD	1-2000 AD
Neolithic Age Begins	Bronze Age Begins	Iron Age Begins	Roman Period	Written History

If my hypothesis is correct, these post holes housed posts which had no ritual connotations, as associated with 'totem poles', but were in fact functional mooring posts for boats. They were utilised for unloading cargo, and moreover, they could be used as simple lifting devices, created by placing a similar sized cross beam across their top, through use of a simple mortise joints, as we see clearly on Stonehenge lintels.

This lifting device could have been used to raise stones from boats during high tide by simply tying the stones to the cross beams. As the tide receded, the boat holding the stone would naturally lower in the water, lifting the stone 'like magic' into the air. This is the first example of a hydraulic lift, which shows the level of sophistication in our ancestors thinking. The stone could then be lowered to either a sledge or rollers placed under the cross beam for the 50m journey to the top of the hill.

NB. At the time of print of this second edition, this process has been recreated by the Discovery Channel in the UK at Poppit Sands, in Cardigan, West Wales for a film to be shown in 2013.

My hypothesis allows us to now construct a map of Britain during Mesolithic times. Taking for granted that Britain consisted of a series of smaller islands and large waterways. Therefore, a more direct path existed from the Preseli Mountains in Wales to the shores of Stonehenge.

Archaeologists have always calculated that in order to obtain the Bluestones, our ancestors had taken a very long and dangerous boat trip around the coast of South West Britain to an outlet in the South that allowed boats to travel up the River Avon or to the banks of Somerset in the North, and had then dragged the stones some 50 miles South, to their resting place in Salisbury Plain.

Proof of Hypothesis No. 18

The increased water tables in Mesolithic Britain would have made it a far easier task to move the Bluestones by boat, via a direct water route from the Preseli Mountains in Wales.

Clearly my hypothesis now enables us to understand how they could have easily travelled on the direct water route between South Wales and Somerset to bring these stones to Stonehenge and how easy it was for them using tidal hydraulics to place these stones carefully on to boats for the journey. To confirm our theory, another post hole was found in an extended part of the visitors' centre in 1989 by the company, Wessex Archaeology. The dating of this post hole proves the 1966 dates to be accurate.

But before we look at this date a more important point should be made about Stonehenge, archaeology and the science of ignorance, as the findings of this book have been available to anyone with an enquiring mind to find. Yet the establishment maintains a false story of our ancient history and this can be seen by its traditional and out dated conventions of dating sites. To illustrate this let's look at what happened when these post holes were found and how the establishment has and still is trying to resist change.

CASE STUDY - An inconvenient truth

In 1935 a car park was created to the north of the A344 road from Amesbury to Devizes, some 100m north-west of Stonehenge, so that visitors' cars would not hinder traffic. Although the area stripped of turf and soil was examined by W.E.V. Young, an experienced excavator who worked with Alexander Keiller at Avebury, no features were found. The car park was later doubled in size but as the experts assumed that no activity would be in that area, there was no record of archaeological work undertaken.

Q. Why did Young not notice four, one metre post holes - not exactly small are they?

Q. If four massive features were 'missed' what other features were 'overlooked' or destroyed?

Q. If W.E.V. Young was just a 'poor' archaeologist, why did he become the curator of the Avebury Museum?

However, in 1966 when it was again extended (so that it was by this time four times the size of the original), a series of circular features cut into the chalk bedrock, and set roughly in line, were observed. Excavation was undertaken by Faith Vatcher, at that time the much-respected curator of the Avebury Museum, with her husband Major Lance Vatcher.

They discovered that three of the features were substantial post-holes cut into the ground, while a fourth was the place where a tree had once stood. Whereas tree-throws are commonly found on the chalk of this area, these pits were more unusual. There is no record of whether the area where the new visitor facilities were to be placed was investigated by the Vatchers. However, when these facilities were also enlarged in 1988, a fourth post hole was discovered.

timeline

4500 BCE	2500 BCE	800 BCE	0-400 AD	1-2000 AD
Neolithic Age Begins	Bronze Age Begins	Iron Age Begins	Roman Period	Written History

Ronald Hutton (in 'Pagan Religions of the Ancient British Isles' 1991) mentions that neither Stonehenge nor Mesolithic experts took much interest in the discovery, maybe because it didn't fit into their ideas of how things had been in the area. This is probably why after taken samples of the Post holes the samples were sealed up and sent into an achieve without being carbon dated to verify the Late Neolithic date they suggested in the report.

Q. Why did Vatchers not send away the samples for carbon dating?

Q. If the fourth pit was a tree hole and this area was a forest, where are the others in the car park?

Q. If they found three pits in a row, why did they not follow the line and find the post hole found in 1988?

These were skilled archaeologists that went on to further their career excavating other sites, writing books and lecturing. The mystery of the post-holes would have remained hidden away if it was not for the enquiring mind of a graduate Susan Limbrey, who realised that Pine wood would not be growing in the Boreal Period of Stonehenge. At that point in 1975 two of the three carbon samples were carbon dated.

Pit A   HAR-455 9130+/-180 BP   corrected to   8820 BCE to 7730 BCE

Pit B   HAR-456 8090+/-140 BP   corrected to   7480 BCE to 6590 BCE

Q. When they found the wood was Pine, why did they not know that Pine only grew in this area in the Mesolithic?

Q. Why were only 2 of the 3 samples carbon dated?

Q. Once the date was verified why was the car park not closed and dug up to see what else was missed?

Thirteen Years later a fourth post hole was discovered by Martin Trott, a young graduate of Southampton University who at the time was working for Wessex Archaeology (later he joined the Inland Revenue.) The four pits were all roughly circular, 1.3m to 1.9m in diameter, and 1.3m to 1.5m deep, and appear once to have held substantial posts 0.75m in diameter. Unlike the Vatcher's pits, Trott's was found to have been re-cut at a much later stage and subsequently deliberately back filled.

Q. Did the other three pits have this detail and was not accurately reported?

17000 BCE	**10000** BCE	You are **HERE** ◆
Ice Age Ends	Mesolithic Age Begins	

Only a few undiagnosed flint flakes and other remains were incorporated into the fill of the pits. However, charcoal from the fills was found to be solely from pine trees. More surprisingly, radiocarbon dates obtained from this charcoal suggest that the pits had been dug between 8090 BCE to 7090 BCE.

Post Hole 9580 QxA-4220 8400+/- 100 BP corrected to 7580-7090 BCE

QxA-4219 8520+/- 80 BP   corrected to 7700-7420BCE

GU-5109 8880+/- 80 BP   corrected to 8090-7690BCE

Q. As this find would verify the astonishing finds of the 1960's, why was a junior member of the archaeology community responsible for the excavation?

Q. This post hole was used for over 400 years, so where is the evidence of settlements (houses) in this area?

Q. To cut down a tree then dig a hole then stand the post upright using stone tools would have taken weeks - where are the camp fires and stone tool flints from sharpening the axes?

According to the book - Stonehenge in its landscape; Snail shells and pollen grains preserved in the fill suggest that the pits were cut in woodland. Taken together, the disposition of soil layers within the pits, the pine charcoal, radiocarbon dates and snail shells provide sound evidence that during the Early Mesolithic a group of people erected stout pine posts in the middle of a mature pine and hazel (Boreal) woodland.... "They are likely to be individual uprights, perhaps reminiscent of those American Indian (totem poles)"(Cleal et al 1995, 43-56).

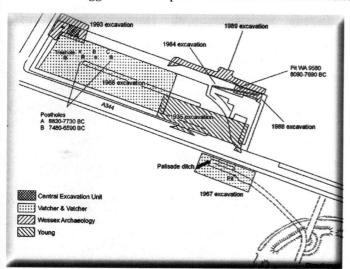

Although Mesolithic sites normally produce evidence for light structures, such as shelters, huts or hearths, the lack of these features in the Stonehenge Car Park is without parallel.

Q. American Indian 'totem poles' are placed on a plain so they can be seen from a distance - why were they buried in a forest?

timeline

4500 BCE	2500 BCE	800 BCE	0-400 AD	1-2000 AD
Neolithic Age Begins	Bronze Age Begins	Iron Age Begins	Roman Period	Written History

Q. Why place the 'totem poles' in a valley when there is high ground just 50 metres away?

And if we use the average date for the construction of these post holes the frequency and alignment just makes the current explanation of these features a complete nonsense. The 'experts' would have you believe that: Post A would be planted in about 8275 BCE - then some 385 years later they planted WA 9580 (75 metres away), returning 855 years later Post B is then planted next to Post A in an alignment with Post C (who's date may never be known), leaving no trace of occupancy as this is a traditional hunter-gather ceremonial meeting spot.

Q. Pine is softwood and would rot in less than 25 years, how did they know where to plant the second, third and forth post?

Q. If they kept coming back year after year to the same spot, why is there no evidence in the car park unlike other sites?

Q. As the post had rotted, why cut a new hole, why not use the old hole?

The 'experts' would have you believe that the same type of hunter-gathers returned to the same spot some 4,000 years later (as the post holes are not dated beyond 7000 BCE) by complete chance (as all the totem poles had rotted away) to build the first phase of Stonehenge!!

Clearly this version of events just does not stand up to scrutiny. The history of the site including the original discovery shows that an attempt has been made to 'brush the evidence under the carpet' or to dismiss the findings as superficial to the story of Stonehenge, as any other conclusion will embarrass the 'experts' and English Heritable that has spend so much time and money convincing the public that they known exactly what happened in prehistoric times at Stonehenge.

Proof of Hypothesis No. 19

The post holes found in the car park at Stonehenge have been dated between 8500 BCE to 7500 BCE. These holes lay on the exact position and height of the prehistoric low tide shoreline. This not only proves existence of water but also proves the date of Stonehenge's first construction.

17000 *BCE*   **10000** *BCE*   You are HERE

Ice Age Ends   Mesolithic Age Begins

The one thousand year variation in dates could be due to the posts being replaced as 'the mooring station' over a period of time. Of course, this would have varied from post to post depending on wear and tear and only fragments of the very final post in each hold would have remained - hence explaining the variation of dates.

Pit 9580 (as it's known in archaeological circles) was found at the time to be 1.9m in diameter and 1.3m deep and it was calculated that the original hole would have been 1m in diameter and 1.3 m deep - the exact same dimensions as the 3 other post holes found in the car park in 1966.

This post hole was later re-cut and widened to the 1.9m. But the most interesting find in the entire site is this additional post hole discovered 1989 by Wessex Archaeology. A piece of Rhyolite (62g) was discovered in the hole at 20cm below the surface. Now, on the face of it, this doesn't sound like an earth shattering discovery, but when you realise that Rhyolite is the technical term for BLUESTONE, it takes on a whole new meaning.

The soil above and below the Rhyolite/Bluestone was carbon dated and it proved that Rhyolite/ Bluestones were at Stonehenge between 7560 BCE and 7335 BCE.

As previously shown, given that Bluestones were only found in South Wales at the time, they would have to have been transported by water and as also established earlier in this chapter, this means that once the Bluestones arrived at their destination, the posts were used to unload them. This then, gives us the original construction date of Stonehenge NO LATER THAN 7560 BCE to 7335 BCE.

Proof of Hypothesis No. 20

The piece of Rhyolite/Bluestone discovered in the 1989 excavated post hole, has been dated between 7560 BCE to 7335 BCE, proving not only that the original post hole is older than archaeologists believe, and more importantly, offers the first real date of when the Bluestones arrived at Stonehenge and construction began.

timeline

4500 BCE	2500 BCE	800 BCE	0-400 AD	1-2000 AD
Neolithic Age Begins	Bronze Age Begins	Iron Age Begins	Roman Period	Written History

Chapter 9 - The Lost North West Entrance

This chapter will provide evidence of a lost entrance at Stonehenge and will show that although used as a spa, the original site was also utilised as a mortuary and was orientated towards the moon and its phases.

North West Passage

Based on my hypothesis, there should be some evidence of a 'lost' processional walkway on the North West side of the monument, towards the original ancient shoreline and mooring station, which would have acted as a natural entry route to the site. Archaeologists have discovered a series of holes (known as Q & R holes) in a semi-circular shape at the centre of Stonehenge, which predate the current Sarsen Stones. The most interesting aspect of this semi-circle is its alignment in a North Western direction.

In astronomical terms this is very important as it faces the general direction of the midsummer moonset - highlighting the relationship of the moon to the site, as opposed to the sun as many 'experts' would have us believe. This connection is reinforced as the sites construction is based on the tidal waters which, in turn, are dependent on the moons position reflected by the location of the Bluestone marker that would have been present sitting in the Aubrey holes. These markers could forecast the daily and seasonal tides which, as a result, would fill the moat with water. Consequently, the importance of the moon is linked with the theory that the original purpose of Stonehenge was as a monument to the dead. This mythology can even be seen today as death and ghosts are associated with the same symbolic objects such as the moon, night-time and gravestones.

One of the greatest archaeological mysteries of the Mesolithic and Neolithic periods is associated human remains.

We know that the population would have been quite small during these periods in history, but the lack of grave plots, has forced archaeologists to investigate non-burial alternatives. One of the favoured alternative methods to burial is known as 'excarnation'. This process, was commonly practiced by Native American Indian tribes, it involves the body being left open to the elements to decompose – a process of returning the body to mother earth. The larger bones would then be collected together and buried together with other members of the tribe or family in tombs. We know that such groups of bones have been found in the Long Barrows surrounding Stonehenge, but to date, no dedicated excarnation sites have been found, until now.

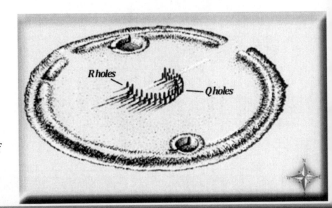

Looking at these Q & R post holes, these form a pre-Sarsen semi-circle at Stonehenge. If you look closely, they consist of two parallel sets. Moreover, they emulate not a 'horseshoe' as archaeologist's normal refer to such objects, but a

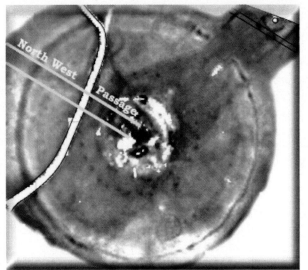

Resistance survey shown an ancient path

over the moat to the north west

'crescent moon' for obvious reasons, in preparation for the journey to the land of the dead. Another point of interest is the distance between each stone in each individual set, which is approximately 2 metres. The 2 metre (6 foot) spacing of these stone holes should seem familiar as it's the same length as an average coffin or grave pit. As this was a monument to the dead, it would seem appropriate that it would be designed in such a way as to allow bodies to be positioned at its centre, aligned with the setting moon.

It's likely that the platforms upon which the dead were laid were either wooden or stone.

Moreover, a 'resistance survey' taken in May 1994 clearly shows a path leading from the centre of the monument to the car park post holes, past the expected location of Aubrey Hole 40. The lack of evidence for this post hole clearly indicates that there once existed a pathway that passed to the North West on the original plan of the site.

With the nondescript name of WA 9421 or the Palisade Ditch, yet another mysterious even lesser publicised feature has been found which does not fit the current expert theories. This Palisade ditch lies in the North West part of the site, between the Stonehenge monument and the visitor's car park and runs across the site running in a South West to North East direction. It spreads from Stonehenge Bottom to Stonehenge Down. From

Proof of Hypothesis No. 21

The existence of the R & Q post holes, forming the semi-circle, pointing in a North West direction and both the geophysical and missing post hole evidence, proves that the original monument was orientated towards North West and the mid-summer moon setting. Consequently, the path from the centre of the Bluestone crescent moon leading to the landing site and our hypothesis' Mesolithic shoreline.

timeline

4500 BCE	2500 BCE	800 BCE	0-400 AD	1-2000 AD
Neolithic Age Begins	Bronze Age Begins	Iron Age Begins	Roman Period	Written History

the depth of the post holes it has been calculated, by archaeologists, to be quite high and would created the ideal barrier or entrance to the monument, as the interior would have been hidden to outsiders on the land until entering through the Palisade.

As my hypothesis has shown that three sides of the monument were cut off by water, the palisade would have effectively created a barrier between two bodies of water cutting off the entire monument, apt for a sacred site dedicated to the dead.

Moreover, to keep wild animals from eating the corpses laid out on the slabs, the palisade extends down the water's edge to the North West and was later extended to the Avenue when the waters fell during the Neolithic Period. This clearly indicates that the use of the site as an excarnation facility must have stayed relevant for over 5,000 years, as they increased the palisades length to match the water's falling levels.

The palisade cuts off not only the peninsula from the mainland but encloses a second peninsular containing some of the most important Long Barrows at Stonehenge - adding credibility to the belief that the site was created for the sick and the dead who were then buried on the same isolated island peninsular if the treatment was unsuccessful.

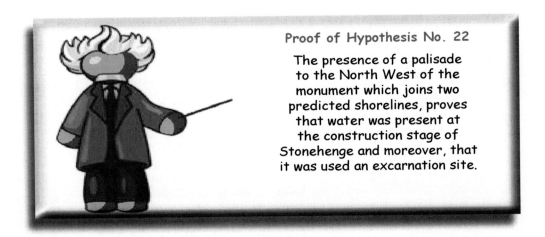

Proof of Hypothesis No. 22

The presence of a palisade to the North West of the monument which joins two predicted shorelines, proves that water was present at the construction stage of Stonehenge and moreover, that it was used an excarnation site.

Aubrey Holes

Gerald Hawkins, an American astronomer, published the results of an intense study of Stonehenge's astronomical alignments in Nature in 1963. In the article he described how he had used a computer to prove that alignments between Stonehenge and 12 major solar and lunar events were extremely unlikely to have been a coincidence (Castleden, 1993).

His book, Stonehenge Decoded, containing the fully developed theory, appeared in Britain in 1966. He described how he had found astronomical alignments among 165 points of Stonehenge, associated purely with the Sun and the Moon, and not with any stars or the five naked-eye planets (Mercury, Venus, Mars, Jupiter and Saturn).

Moreover, he discovered that lunar eclipses could be predicted through a system of moving markers around the circle of Aubrey Holes (a series of 56 holes situated around the Stonehenge moat, which archaeologists believed originally held the Bluestones).

Anyone who has ever tried to make a model of how the Sun and Moon move around the Zodiac will end up, most likely, with a circle of 28 markers around a central earth. Moving a 'Moon-marker' one position per day and a 'Sun-marker' once every 13 days provides a calendar, accurate to 98%.

Every year, for about 34 days, the full and new moons occur near the Sun's path (the ecliptic) and eclipses result. These two occasions, 173 days apart, move backwards around the calendar taking 18.6 years to complete a revolution. The precise two points where the moon crosses the apparent path of the sun through the zodiac (the ecliptic) are the lunar nodes, as mentioned above.

By doubling the sun-moon calendar to 56 markers, we can obtain an accuracy of 99.8% and meet the handy convenience that 18.6 x 3 is almost the same as 28 x 2. Now, a 3:2 ratio enables eclipses to be predicted to high accuracy.

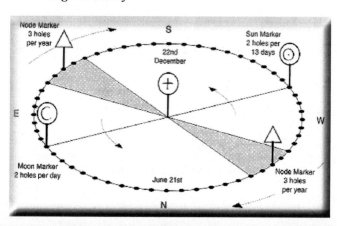

This complex computation could be calculated through the 'Aubrey Holes' features at Stonehenge. By simply moving one marker, once a day and another marker every 13 days, not only could the spring high tides (by when the marker reaches a particular stone) be predicted, but the Solar and Luna eclipses could also be calculated with great accuracy – a stunning achievement for so the called primitive Stone Age Man and probably beyond most readers of this book, even with our cherished education system.

timeline

4500 BCE	2500 BCE	800 BCE	0-400 AD	1-2000 AD
Neolithic Age Begins	Bronze Age Begins	Iron Age Begins	Roman Period	Written History

Chapter 10 – The Avenue to a New Neolithic Mooring Place

When the waters started to subside, the builders of Stonehenge had an engineering problem: how to access the falling waters in order to fill the Stonehenge moat, in this chapter we find out how this was achieved. Firstly, by lining the Stonehenge moat with a liner, this was found by Hawley during his excavations in the 1920s and then by building a new earthwork called 'the Avenue' to the monument.

Processional Walkway

The Avenue was created some time after the original moated henge, according to my calculations, once the original Mesolithic groundwater and shoreline lowered towards the new Neolithic groundwater level. When the waters lowered, the builders faced two problems; firstly, the original mooring could no longer accept boats or cargo, so a new entrance was required. Secondly, the moat would no longer fill as it did in Mesolithic times, as the water table had dropped by about 10 metres.

As we have seen from the excavations by Hawley in the 1920s, the builders had added a liner similar to those found in Mesolithic period constructions of 'dew ponds', found in this area. The reason for the liner would have been to retain the water that accumulated from the natural rain and seasonal high tides that would have replenished the moat. It is quite plausible that this insufficient, or infrequently sufficient, water level was a common occurrence in the past, so to carry on the bluestone treatments an alternative method to fill the moat needed to be found.

The Avenue is quite curious and excavations have revealed its development in the past, as indicated by the post holes dotted down its entire length. This clearly shows that it 'adapted' over a long period to match the retreating shoreline in the Neolithic period, which lasted over two and a half thousand years. Moreover, the physical construction of the Avenue earthwork shows the incorporation of water features.

The Southern ditch is much shallower than the Northern ditch of the Avenue. This is because the water lies on the Southern side of the Avenue and does not need extra depth to fill the Southern ditch evenly, while the Northern ditch is further away from the water (and on a gradient). For it to fill evenly to the same depth of water as the Southern ditch, the Northern ditch would therefore need to be deeper.

The wooden poles within the Avenue seem to be rather sporadic and impractical at first glance. But if you then add the prehistoric waters to the Avenue, you can see that the poles actually line up in pairs. These pairs of posts seem to appear every ten metres as you proceed down the

Neolithic Walkway to the new shoreline

Stonehenge in the neolithic period showing new water levels and mooring posts

Avenue, indicating that the Avenue was created in sections. This is a consequence of the periodic need to replace the poles as the shoreline retreated. The first main brace of these shoreline poles are seen quite close to the famous North-Eastern entrance by the Heel Stone, extending to the last evidence of the poles at the 'elbow' at the end.

The current 'expert' theories suggest that the path of the Avenue is a processional walkway down to the River Avon. But careful study of this walkway shows that it makes absolutely no sense in the route it follows - it's not direct, nor logical in its course. What the Avenue shows as we see today, is the continuation of its association with the River Avon, after the original Avenue that finished at Stonehenge Bottom has fell into disuse. This explains the strange haphazard route it now shows. Archaeologists, not finding the true reason for its course, attempt to justify the route as a 'ceremonial' pathway, which ignores the facts. These 'facts' are compounded as later ancestors constructed Barrows that mimicked their ancestors constructions and placed burials within them, totally confusing the timeline.

The later attempts to keep connection to the original builders of Stonehenge can be seen in the Avenue as it moves off to the North East for about 500 metres (the original construction), then suddenly swings off East for another 1,000 metres, then turns again and heads South, like a stretched lower case 'n'. Archaeologists suggest that this is a 'natural route' deliberately passing burial mounds that lay to the East of Stonehenge.

But haven't they put the cart before the horse on this one?

Proof of Hypothesis No. 23

The variation in the size of the ditches either side of the Avenue, to allow an even depth of water on both sides, shows that water was present at Stonehenge in Neolithic Times.

timeline

4500 BCE	2500 BCE	800 BCE	0-400 AD	1-2000 AD
You are HERE				
Neolithic Age Begins	Bronze Age Begins	Iron Age Begins	Roman Period	Written History

The facts are; the walkway came first, followed by the introduction of the burial mounds – which is borne out by carbon dating showing that the Avenue was first introduced in Neolithic times, and the burial mounds much later in the Iron Age. Also archaeologists have identified that the Avenue was built in sections; the first section up to the 'elbow', according to Julian Richards in his book 'The Stonehenge Environs Project', was built during what archaeologists call 'Period II', including:

- **Modification of the original enclosure**
- **Entrance**
- **Construction of the first straight stage of the Avenue**
- **Erection of the Station Stones**
- **Resetting of the two entrance stones**
- **Dismantling of the double bluestone circle**

If the Avenue was a processional route, why stop it in the middle of nowhere, only to eventually be joined to the Avon some thousands of years later?

Our hypothesis (unlike the present expert views) answers this question fully. Clearly, the Avenue was built to meet the new shoreline that emerged during the Neolithic period, where (as at the old North West entrance) they received and moored the boats, cargo, and even the stones. Consequently, this then acted as a new entrance to the Stonehenge monument.

Proof of Hypothesis No. 24

According to 'The Environs Stonehenge Project', the Avenue terminated at the 'elbow' of the processional causeway in an abrupt ending, without reason. Our hypothesis explains that the reason for the termination was the predicted shoreline of Stonehenge during the Neolithic Period.

17000 BCE

Ice Age Ends

10000 BCE

Mesolithic Age Begins

Further evidence of this shoreline can be seen in the post holes found in the Avenue. These post holes do not make sense, as they would have blocked the walkway, unless they are mooring posts for various phases in the Avenue's construction and water level. A small survey carried out from 1988 to 1990 investigated the last 100m of the Avenue up to the elbow. It found 14 large post holes that could have been used as moorings for boats. These moorings are paired off at 45 degrees to the Avenue, which would match the exact shoreline predicted over several periods.

We can also date the extension past the 'elbow', as it no longer has the elaborate moats on either side of the processional walkway. The Avenue beyond the 'elbow' travels over a series of hillocks that could not hold water at any level, and therefore made such moats useless and unnecessary. This, again, shows that our ancestors were practical and pragmatic people who included aspects such as ditches for useful purposes, not for religious or aesthetic reasons as current archaeologists believe.

At this stage of the monument's construction, I believe Stonehenge finally turned from being a monument to the moon, for curing the sick and excarnating the dead, to being a shrine to life and sunrise, which was in the Bronze Age and thereafter, taken over by a new civilisation of druids and tribal warriors.

Proof of Hypothesis No. 25

Excavations undertaken between 1988 and 1990 in the 'elbow' part of the Avenue, show that 14 large post holes were positioned at an angle that would meet the predicted shoreline of my hypothesis during the Neolithic Period.

timeline

| 4500 BCE | You are HERE | 2500 BCE | 800 BCE | 0-400 AD | 1-2000 AD |
| Neolithic Age Begins | | Bronze Age Begins | Iron Age Begins | Roman Period | Written History |

Chapter 11 - Snails can tell us an old story

In this chapter, we look at a method that archaeologists use in trying to understanding how past communities lived amongst the monuments they built. One method used allows us to understand what kind of landscape they inhabited. The rolling landscape, vegetation and farming patterns we see today are largely a product of the modern agriculture, although much of it has its heritage in our ancestors' landscape of six to ten thousand years ago. This landscape is not the one that our prehistoric ancestors would remotely recognise. Only the general shape of the landscape remains unchanged – the rivers, climate, vegetation, wild animals, and patterns of land use have all been completely transformed by our dry climate and industrialisation.

Archaeologists have recently used the discovery of small snails in the excavation soils in an attempt to understand the prehistoric conditions, and possible uses, of the land surrounding archaeological sites. Scientists believe that if we can understand the species and number of these creatures found within the soil samples excavated, we may have a clearer indication of the environment our ancestors inhabited.

Mollusca

If we look at the site at W58 Amesbury(Coneybury Henge), archaeologists (in this case Roy Entwistle for the 'The Stonehenge Environs Project') traditionally interpret the evidence gained by identifying individual mollusc species as indicating that taller vegetation must have been present nearby at the time of construction - as the C. tridentatum snail inhabits only grasslands. Unfortunately, when you look at the numbers, the assumption just does not 'add up', for that breed of snail numbers only 204 of the total of 1145 snails found in that sample. Some open-field snails, like C. Tridentatum, also seem to love shaded habitats such as deep forest litter and primeval woodland, according to Dr Roy Anderson's report (Species Inventory of Northern Ireland, 1996). Consequently, the only thing we know for sure is that snails like dark wet places, so they can be found under logs; but if there is a lack of trees, they favour rocks.

This information can be reinterpreted as the tendency for snails to like rocks which is good news for us, as we can now show with a great degree of certainty when the Monolith stones (if any) were erected, because there seems to be a massive increase in the total snail numbers at a soil depth at a depth of 1.2m to 1.3m. What needs to be kept in mind is that the Mollusca are aquatic animals and can only survive in wet habitats, with certain plants. The more damp, the better for the humble snail; it does not thrive on dry, barren land. I believe that the

W58 Amesbury 42 Long Barrow	
Depth (m)	Numbers per Kg
0 - 0.15	45
0.35 - 0.45	145
0.6 - 0.7	238
0.7 - 0.8	206
0.85 - 0.95	636
1.2 - 1.3	1145
1.45 - 1.55	473
1.65 - 1.75	294
1.85 - 2.00	148

increase in snail numbers in the past not only shows the type of vegetation in the past but, more importantly, the climate and soil conditions - in particular, the amount of water in the environment.

The snail counts for Coneybury Henge also show that the ditch was at one stage much wetter, at 1.0 m – 1.6 m excavation depth. There are 2,200% more snails present in the ditch than today, which shows either:

- **The flora had overgrown the site**

- **The stones (a snail's perfect home) were present at the site**

- **The ground was wet and boggy, after filling with water and then silting up**

Moreover, it is shown that at 1.8 m to 2.8 m depth there were no snails at all – this could only happen if either the landscape turned into a desert, or the ditch filled with water so that the only soil floating to the bottom of the ditch was water silt. When you realise that from 1.8 m to 2.2 m the ditch was filled with silt from water, the sequence suddenly makes perfect sense:

W2 Coney Henge Ditch Segment	
Depth (m)	Numbers per Kg
0 - 0.2	130
0.2 - 0.4	103
0.4 - 0.6	228
0.6 - 0.8	455
0.8 - 1.0	2587
1.0 - 1.2	2896
1.2 - 1.4	1206
1.4 - 1.6	1728
1.6 - 1.8	591
1.8 - 2.0	0
2.0 - 2.2	0
2.2 - 2.4	0
2.4 - 2.6	Data Lost
2.6 - 2.8	2
2.8 - 3.0	63

- **A depth of 3 metres or more correlates to the end of the Ice Age, when snails lived in small numbers on tundra (treeless plains) and sparse grassland.**

- **At 2.8 m to 1.8 m, a ditch was built and water filled the ditch – a few snails falling into the water would account for the small numbers found.**

- **At 1.8 m to 0.8 m, the ditch dried up and became a marsh surrounded with stones, a perfect environment for our aquatic snail to live and breed within.**

- **At 0.8 m to 0.4 m, the stones of the henge were removed and the ditch became a grassy dip in the ground.**

The most amazing thing about this site can be seen in the cross-section of Coneybury Henge. It clearly shows why this site was chosen as a henge, and why the ditch filled with water. During the Mesolithic Period, water would have been present on two sides of the site. In fact, the henge sat on a peninsula that overlooked Stonehenge, separated from it by a vast waterway over 100m wide. Moreover, if you look at the Stonehenge site in comparison, you will see that Coneybury Henge is just 3 metres higher. This height difference could be for one of two possible reasons:

- **The groundwater table on this side of the river may be 2 metres higher, which is quite possible as chalk contains strata that can allow this.**

- **The site was built prior to Stonehenge, when the waters were even higher – which makes it a greater shame that farmers have ploughed it into**

timeline

4500 BCE	2500 BCE	800 BCE	0-400 AD	1-2000 AD
eolithic Age Begins	Bronze Age Begins	Iron Age Begins	Roman Period	Written History

complete extinction with the possible losses of its history for all time.

This evidence in itself is quite compelling but, interestingly, like the Stonehenge ditch, Coneybury Henge's ditch contained dark material showing that water was present when the ditch was constructed. If we look at a soil report for the same ditch where the snail survey was completed, Helen Keeley found 'a turf line' of soil between 2.78 cm and 2.81 cm composed of humic (organic) materials.

Coneybury Henge

So, is this the same dark material Hawley found at Stonehenge?

I believe so; in fact, within this layer Keeley also found the largest deposit of calcium carbonate in the entire sample. Calcium carbonate is a common substance, present in rocks and many other objects all over the world; it is the main ingredient of the shells of aquatic organisms and snails. Calcium carbonate producing creatures such as corals, algae, and microorganisms usually live in shallow water environments; because they need sunlight in order to make calcium carbonate.

It was also found throughout the rest of the sample, but in much smaller amounts. The greatest amount was left at the bottom – clear evidence that shortly after the time of its construction, Coneybury Henge ditch was full of water, just like Stonehenge. In conclusion, we have now shown that Mollusca data can be very useful to indicate; when the flood waters started at the site, when the stone monuments were erected, when the groundwater table dropped on the site when the forest clearances started so that the site could be accessed by land.

Proof of Hypothesis No. 26

The lack of Mollusca and the finding of calcium carbonate, at certain levels in the ditch of Coneybury Henge prove that the moat surrounding the monument was full of water due to the high water table predicted during the Mesolithic Period by my hypothesis.

17000 BCE

10000 BCE

You are HERE

Ice Age Ends

Mesolithic Age Begins

Chapter 12 – Barrows – Long and Round

A tumulus (plural tumuli) the word tumulus is Latin for 'mound' or 'small hill', from the PIE root *teuh- with extended zero grade *tum-, 'to bulge, swell' also found in tumor, thumb, thigh and thousand - A mound of earth and stones raised over a grave or graves. Tumuli are also known as barrows, burial mounds, Hügelgrab or kurgans, and can be found throughout much of the world. A tumulus composed largely or entirely of stones is usually referred to as a cairn. A long barrow is a long tumulus, usually for numbers of burials. The method of inhumation may involve a dolmen, a cyst, a mortuary enclosure, a mortuary house or a chamber tomb.

Not overly clear, is it!

It's a grave, although the Latin word does not mean grave, but a 'small hill'. That's a big difference; they also can be long or round, stone or earth, etc. When we look at the traditional archaeological definition of tumuli, we are offered:

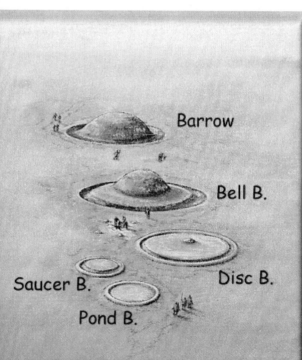

- **Bank barrow**

- **Bell barrow**

- **Bowl barrow**

- **D-shaped barrow,** round barrow with a purposely flat edge at one side often defined by stone slabs

- **Fancy barrow,** generic term for any Bronze Age barrows more elaborate than a simple hemispherical shape.

- **Long barrow**

- **Oval barrow,** a Neolithic long barrow consisting of an elliptical, rather than rectangular or trapezoidal, mound.

- **Platform barrow,** the least common of the recognised types of round barrow, consisting of a flat, wide circular mound, which may be surrounded by a ditch. They occur widely across Southern England with a marked concentration in East and West Sussex.

- **Pond barrow,** a barrow consisting of a shallow circular depression, surrounded by a bank running around the rim of the depression.

timeline

4500 BCE	2500 BCE	800 BCE	0-400 AD	1-2000 AD
Neolithic Age Begins	Bronze Age Begins	Iron Age Begins	Roman Period	Written History

- **Ring barrow,** a bank which encircles a number of burials.

- **Round barrow,** a circular feature created by the Bronze Age peoples of Britain and also the later Romans, Vikings, and Saxons. Divided into subclasses such as saucer and bell barrow. The Six Hills are a rare Roman example.

- **Saucer barrow,** circular Bronze Age barrow featuring a low, wide mound surrounded by a ditch, which may be accompanied by an external bank.

- **Square barrow,** burial site, usually of Iron Age date, consisting of a small, square, ditched enclosure surrounding a central burial, which may also have been covered by a mound

Prehistoric
sites are
83% up a
hill!

I make those 13 different types of barrow!

If the archaeologists are right, and barrows are just grave plots, they could be any shape. But do we really believe each had a different function or the same purpose in various forms? The truth is that there are (in my view) just five categories of barrow:

- **Long Barrow** - the biggest and earliest form used for burials

- **Round Barrows** – big and round used as markers

- **Pond Barrows** – wells for water extraction

- **Disturbed barrows** – Round/Long/Pond Barrows that have degraded over 5,000 years by the elements and man's attempts to destroy or excavate them

- **Copy barrows** – imitation Barrows from a later date, mimicking their ancestors.

Barrow Altitude

We have taken a sample of 50 prehistoric sites around and including Stonehenge, to look at the topology of the area and their location. Our findings clearly show that the traditional belief that prehistoric man made his burial and ceremonial structures on top of hills is wholly incorrect. In fact, in our survey, only 8% of sites are on top of a hill.

On average, sites are located 83% of the way up a hill.

This includes our most ancient site, Stonehenge; we are led to believe that our ancestors brought the bluestones all the way from Wales - some 250 miles - only to stop 50 metres short of the top of the hill because they were..... tired? Or some other obscure unknown ceremonial or astronomical reason.

As we have seen from previous chapters, Long Barrows were created for the bones of the dead after an excarnation – this is when a body is left uncovered and exposed to the elements for birds to pick the skeleton clean. It is believed that the bones were brought on a boat to their final resting place, as all barrows were built in close proximity to a river.

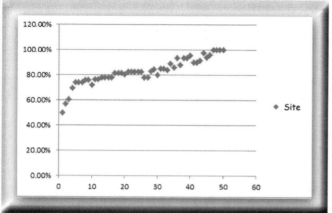

50 Site locations around Stonehenge

I believe that without a rise in groundwater levels, this process would have been totally illogical and in some cases impossible, looking at the locations of these sites. As an example, if we look at the nearest Long Barrows to Stonehenge, we find two next to each other at a crossroads called simply 'the Long Barrow Crossroads'. We have seen in a previous chapter on earthworks, these objects had canals built for them. But if we are 'incorrect' and the experts are right then the most likely routes for funeral processions would be from the rivers on either side of the site at their current level. Unfortunately, for these archaeologists, these are all over 3 km away from the Long Barrows.

Moreover, some of these routes pass higher ground on their way to the Long Barrows, 'begging the question', why would you not build a barrow there?

Other routes do eventually reach the top of the hill upon which the Long Barrows are built, but then have to come down again because (as we have shown) the barrows themselves are partway down the slope. Consequently, we are asked by archaeologists to believe that our ancestors took their funerals up and down hills, covering 3 km to bury their dead, when higher sites were available along the processional routes.

My hypothesis makes perfect sense of the locations of these ancient monuments, by understanding where they stood when they were first built – on the shorelines of vast prehistoric waterways. For our survey also shows that these barrows were not constructed at random heights. In the Stonehenge sample of 50 sites, the lowest burial was at 89 m OD (above current sea level), the highest at 115 m OD.

If you were going to go to all the effort of constructing a Long Barrow - which would have taken many months, if not years, working with just stone axes and antler picks - then you can imagine that you might either build it at the top of a hill to provide the dead with a highly visible monument, or simply avoid all the extra work of carrying things up the slope by burying them at the bottom of the hill.

The only logical reason you would construct your barrow at the mid-point of a hillside would be that there was another overriding factor to

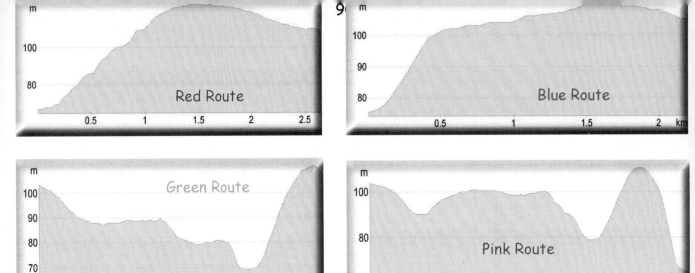

consider. This reason can only be that you can't build a barrow under water, and the groundwater at that time lay between 75 – 90 metres OD.

The variation in this figure is due to the groundwater tables falling from the Mesolithic (high) to the Neolithic (low), after which barrows were no longer built for their original purpose (although Bronze and Iron Age people may have copied their ancestral rituals – but that is not the case in this area).

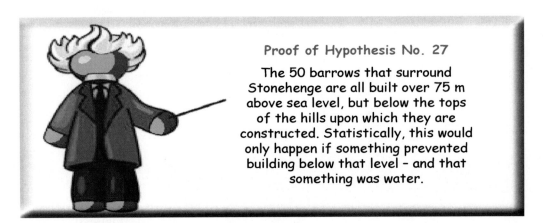

Proof of Hypothesis No. 27

The 50 barrows that surround Stonehenge are all built over 75 m above sea level, but below the tops of the hills upon which they are constructed. Statistically, this would only happen if something prevented building below that level – and that something was water.

Long Barrows

The first thing we should note about Long Barrows is that they are unique to Northern Europe, unlike Round Barrows, which are found all over the world. Archaeologists agree that the Long Barrow is the oldest monument to exist in our landscape (there is even some carbon dating evidence from Carnac in France that dates St Michael's tumuli (a Long Barrow) to be in existence at least by 6850 BCE - which has been dismissed as an anomaly). This belief originates from the fact that the structure is very elaborate and includes megaliths, as seen at Avebury as well as carbon dating of the bones found inside. Moreover, they are also aware that there are bones from many dead people collected together in the chambers of these types of burial mounds, rather than in individual graves, or cremations that were seen at a later date.

The number and condition of these bones show us that they were disarticulated, with only the larger bones and skulls being brought to the sites after death, probably after the bones had been de-fleshed. The reason we are interested in these objects is twofold. We believe their shape and design are of great importance. Firstly, the mounds are long and thin, with the entrance at the end of the mound.

The entire Long Barrow mound originally had a ditch dug completely around its exterior, which starts to relate it to an object used by our ancestors after the great flood.

To understand why they would construct a ditch, we must again look at these monuments in the light of our new discovery of waterways at the time of construction. During the prehistoric period, Long Barrows like West Kennet was on a peninsula, surrounded on three sides by water. This groundwater now gives us a clue as to why the ditches that surrounded the monument were dug, for the bottom of the ditches would have been below the groundwater table at this location.

Proof of Hypothesis No. 28

The positioning of long barrows clearly indicates that they were built at the shorelines of prehistoric rives, because the traditional paths from existing rivers indicate that the processional causeways would have either passed higher ground or overshot the brow of the hill.

timeline

4500 BCE	2500 BCE	800 BCE	0-400 AD	1-2000 AD
Neolithic Age Begins	Bronze Age Begins	Iron Age Begins	Roman Period	Written History

They were, therefore, not ditches but moats.

The raised groundwater levels will also give you an indication of how the gigantic rocks, weighing over 5 tonnes, got to the top of these hills, as the shoreline at West Kennet was only 50 metres down the track from where the Long Barrow is today. Moreover, when you add the increased groundwater level to this site, the monument takes on a new perspective, for it is in the shape of a 'longboat' and, therefore, the moat that surrounds it represents the water.

The Long Barrow represents the boat culture of this ancient society; they lived in boats and so, when they died, they were sent on their last voyage by boat to the afterlife. Even today, we still have a custom of placing money over a dead person's eyes as their fare to be collected by the ferryman. Consequently, this also gives us a fantastic insight into the design of the boats used in this period. This boat looks more like a barge than a canoe, with the back end (the stern) being where they steered the craft with a rudder, which means they used sails, not paddles, for power.

Moreover, a sailed boat would have a greater range than a canoe, which is made only for short distances. The boat's front was built at an angle, not flat; this is a design meant for manoeuvrability, speed and distance, clearly indicating the ancestors' knowledge of engineering.

Consequently, this was NOT a primitive society!

The other noteworthy aspect of the Long Barrow is the construction of the end (stern) of the 'boat'. At West Kennet and other Long Barrows, giant megaliths were used to highlight the entrance to the chambers. They are not necessary to the construction, but they are visible from a couple of miles away. Long Barrows, when first constructed, would have been covered not with grass as today, but with the sub-soil that came from the ditch surrounding the barrow. In the case of West Kennet, it would have been bright white chalk. The most remarkable aspect of Long Barrows is that they have been positioned to present a lengthways 'side view' to the river, so as to be seen clearly from the waterways our hypothesis describes.

We have investigated the eight existing Long Barrows that surround Stonehenge and its sister sites, and

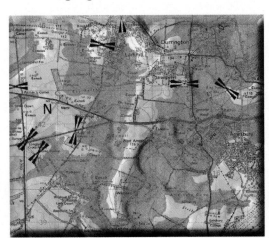

Alignment of Long Barrows around Stonehenge

West Kennet Long Barrow

found that each one points in a different direction as measured by a compass, but all are aligned to the exact contours of the predicted shoreline. Moreover, not only are these 'markers' visible for miles in daylight, but because they are pure white, they can be seen on nights when there is moonshine. Therefore, we can conclude that this society travelled on a 24/7 basis.

We can now also understand why they placed large Mesolithic stones in the stern of the Long Barrow. This would indicate the direction a boat should take if there was a split in the estuary or navigate back from the destination. We have found that the stern stones on all the long barrow boats from Stonehenge to Avebury all point in the direction of Stonehenge.

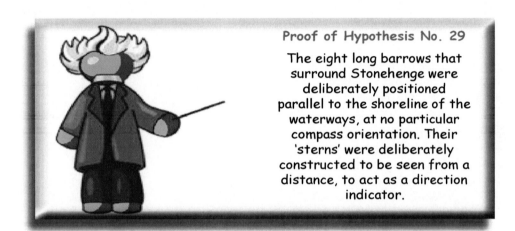

Proof of Hypothesis No. 29

The eight long barrows that surround Stonehenge were deliberately positioned parallel to the shoreline of the waterways, at no particular compass orientation. Their 'sterns' were deliberately constructed to be seen from a distance, to act as a direction indicator.

Round Barrows

In our examination of Long Barrows, we have discovered that this civilisation traded heavily with other civilisations, tribes, clans or groups within prehistoric Britain and beyond. Herefordshire businessman Alfred Watkins was sitting in his car one summer afternoon, during a visit to Blackwardine in Herefordshire in 1921, when he happened to consult a local map and noticed that a number of prehistoric and other ancient sites in the area fell into alignments. Subsequent field and map work convinced him that this pattern was indeed a real one. Watkins came to the conclusion that he was seeing the vestigial traces of old straight tracks laid down in the Neolithic Period, probably, he surmised, for traders' routes. He concluded that after modernisation in the later Bronze and Iron Age periods, the tracks had fallen into disuse during the early historic period. The pattern had been accidentally preserved here and there due to the Christianisation of certain pagan sites that were markers along the old straight tracks. He published these theories in two books.

timeline

4500 BCE	2500 BCE	800 BCE	0-400 AD	1-2000 AD
Neolithic Age Begins	Bronze Age Begins	Iron Age Begins	Roman Period	Written History

Reaction to Watkins' book 'The Old Straight Track' was sharply divided. Many thought he had uncovered a long-forgotten secret within the landscape, and The Straight Track Club was formed to carry out further "ley hunting", while orthodox archaeologists vehemently dismissed the whole notion. And with a few notable exceptions, this situation still exists today. Later in this section we will show that Watkins was right; not only do these tracks actually exist, but there are two kinds of track. First, the older Long Barrows were markers based on islands and peninsulas within the river routes to known sites; then, after the groundwater subsided, Round Barrows were used for overland routes.

When the groundwater subsided after the Neolithic Period, our ancestors needed a further navigational aid to allow them to find the location over land rather than by river as in the past. These markers are known today as Round Barrows and they are consequently, more frequent than Long Barrows because dense foliage makes the line of sight shorter for someone on foot than for someone taking the same route by boat and secondly, there are more routes by foot, than by river to navigate.

The incorrectly conceived aspect of Watkins land markers is that they did not run in straight lines, but to the highest ground. This was because the lower ground that had been flooded in the Mesolithic period would still be boggy and wet in some seasons, and therefore impassable. So these tracks are straight, but not ruler-straight as Watkins first believed.

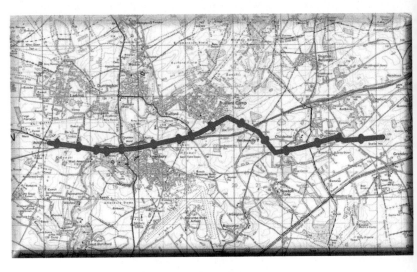

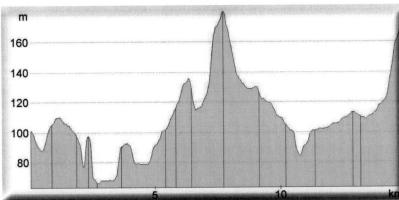

Round Barrow path from Stonehenge to Quarley Hill showing barrow heights and distances apart

One of the closest sites to Stonehenge is Quarley Hill to the east, the path between these two sites passes 13 barrows in a 15 km distance. If you stand on top of one of these barrows, you can see the next one in line very clearly. You must also take into consideration that nearly half the barrows have been ploughed out by farmers, and that the mounds would have been at least 30% larger and pure white in colour at time of construction. Even so, they can still be seen as path markers today in some areas, 5,000 years later.

It is almost impossible to know what information would have greeted a walker in the Neolithic when he reached a Round Barrow, but we still have milestones surviving today and I believe, this is clearly their ancient original form. One can only imagine that somehow the barrow, like a milestone, would give an indication of the distance to be travelled. It should also be noted that the burials within these Round Barrows were placed in them at a later date, which would explain why these barrows do not contain burials, and why at others the dead were buried in strange corners of the barrow and not at the centre as an afterthought.

Pond Barrows

Now we have established the use of the majority of prehistoric barrows by our ancestors, we can look at the 'other' barrows catalogued by archaeologists, to see what function they had in helping our ancestors navigate from town to town in prehistoric days. Pond barrows' shape is just as described: 'pond-like'. And it doesn't take a genius, now that we have proved that groundwater tables were higher in the past than today, to identify what these structures were – artificial ponds.

Water, in any civilisation, is critical to survival. Our ancestors were no exception, so when they travelled on foot to other towns or sites, the provision of water was essential. Most pond barrows have the centre dug out, which would have tapped into the groundwater course; this would allow the pond to flood, depending on the tide levels. This tradition continues into modern days by the use of dew ponds, which are of the same size and shape, but relied more on rain water to fill the pond.

Other Barrows

These barrows are lost, in disrepair, or they are copies from an age after the original barrow builders had left or died.

timeline

4500 BCE	2500 BCE	800 BCE	0-400 AD	1-2000 AD
Neolithic Age Begins	Bronze Age Begins	Iron Age Begins	Roman Period	Written History

Section Three: The Landscape Evidence

The most compelling evidence for my hypothesis - that prehistoric monuments were built around a flooded environment - can be seen within a detailed topological inspection of ancient sites. The landscape layout of these monuments clearly shows that their entrances and mooring areas were oriented to the shoreline contours of their surroundings. At first glance, due to the considerable age of these monuments, it is difficult to see how the landscape originally looked and this is the reason why archaeologists have obviously overlooked the possibility in the past.

However, like jigsaw puzzles, once you have assembled the end pieces and the borders, the picture becomes much easier to understand, and this is the case with our most ancient sites: Stonehenge, Avebury, Old Sarum and Woodhenge. The most effective method we have found for locating this evidence is to look at the profiles of these sites, and then their position relative to the landscape. By doing so, ghosts of the original landscape can be found in the contours that have changed little in the last 10,000 years.

Our research has shown that Stonehenge was not the only monument in this area of the landscape to be affected by the rise in groundwater tables. Three other sites in the general vicinity were similarly affected, all of which indicates that these sites were socially connected by the waterways that once flowed through the area. Consequently, in this section we shall concentrate in detail on these four sites, to allow us to get the best picture on how this civilisation organised itself in this area. It should be remembered that although we are illustrating how the higher groundwater tables affected the Salisbury Plain area, all parts of Britain would have been equally affected. This will allow future investigations to find even more evidence to support our hypothesis.

Chapter 13 – The Stonehenge Landscape
(Case Study No.1)

Looking at the most important prehistoric site in the UK, Stonehenge, we are asked by archaeologists to consider that it is an astronomical calendar showing the rising and setting of the Sun, Moon, etc. This assumption may be partially correct, but why would you place the site halfway up a valley? If you wish to study the stars, or watch the sun rise and fall, with any great accuracy, you will need to locate your site at the highest point available. So why was Stonehenge built where it is, rather than at the top of the hill just 500 metres away that is 30 metres higher?

As there is no observational advantage to placing the site in this location, we must therefore look at how it was constructed, as this may give us a clue.

If we look at a standard Ordnance Survey (OS) map of Stonehenge, it indicates the landscape and topology of the surrounding area by showing the contours of the hillsides. But it does not give you a clear idea of how Stonehenge actually sits in the landscape; for that we need to look at a profile of the area.

As we have already shown, we believe that Stonehenge was built on the shoreline of a vast river complex. If we can see evidence of this river complex on the elevation map, and consequently find Stonehenge is sited on the side of a empty 'Dry River Valley' on its shoreline, we will have yet another proof of our hypothesis.

The first thing you will notice from an elevation map is that Stonehenge is located about two-thirds of the way up the hill upon which it's built. If the archaeologists are correct in their assumption, that it was primarily an astronomical calendar, why wasn't it built on top of the hill just 500 metres away? Especially considering that the builders took the trouble to get stones from 250 miles away in Wales, why would they stop short of the best position?

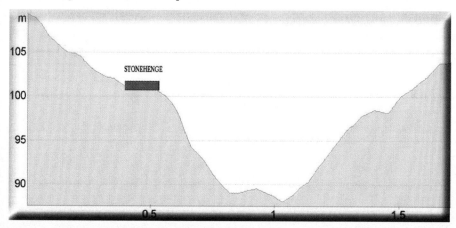

Clearly, the monument's main purpose was not astronomical; or, if it was, there must have been an even better reason for the site's placement within the landscape! Our hypothesis indicates that the groundwater tables during the Mesolithic construction period would have filled the dry river valley with groundwater 30 m above the existing groundwater table – when this happens, the profile changes dramatically.

timeline

4500 BCE	2500 BCE	800 BCE	0-400 AD	1-2000 AD
Neolithic Age Begins	Bronze Age Begins	Iron Age Begins	Roman Period	Written History

And so suddenly, the impossible becomes possible, the implausible becomes credible, for these profiles can only indicate one fact – Stonehenge was built on the side of a hill surrounded by water!

These amazing features are not just found on a single side of the monument; we can go around the whole circle to see these watery features. I believe these profiles tell the entire story of Stonehenge. A picture is said to paint a thousand words; these pictures show that our most famous ancient monument was once a magnificent feature in the landscape on the edge of a peninsula, surrounded by water. Now that we have proved that water existed at the Stonehenge site during Mesolithic and Neolithic times, we can re-sequence the events and building phases by putting together the clues and proofs discussed in the preceding geological and archaeological sections. This will also allow us to place all the information we have gathered into a logical sequence of events.

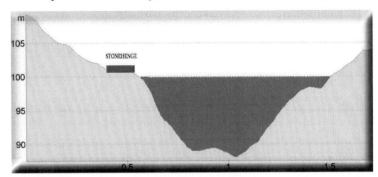

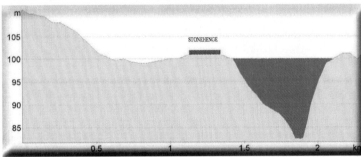

We have split the history of Stonehenge into four separate phases; most of the construction of Stonehenge happened gradually over the years. We know from our own experience of old houses that they start as a specified building, and change with time to become something quite different to their original use – hence all the new pub churches! This is the case with Stonehenge. This will also give us an opportunity to summarise what we have learnt to date about this site.

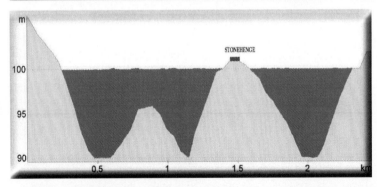

Proof of Hypothesis No.30

The elevation data shows that Stonehenge was placed three-quarters of the way up a hill, which matches the predicted Mesolithic water tables.

Phase	Construction/Use	Date BCE
I	Moat	8500 - 7000
	Bluestones - Aubrey Holes	
	Excarnations	
	Palisade	
II	The Avenue	4000 - 4300
	Sarsen Stones	
	Station & Heel Stone	
	Dry Moat Lining	
	Palisade Extension	
	The Cursus	
III	The Avenue - Extension	3500 - 2500
	Bluehenge	
	Round Barrows	
	Moat cleaned out to become a ditch	

My proposed new time-table for the construction phase at Stonehenge

timeline

4500 BCE	2500 BCE	800 BCE	0-400 AD	1-2000 AD
Neolithic Age Begins	Bronze Age Begins	Iron Age Begins	Roman Period	Written History

PHASE I Construction – 8500 BCE to 7000 BCE

Moat

As we have shown in previous sections, the first phase of Stonehenge was the construction of the moat, which was originally built as individual pits or baths with internal walls, rather than a continuous deep ditch around the edge of the site. We imagine that these pits had either a large bluestone on the floor of the pit, or chips of bluestone added like bath salts, to obtain the full benefit of the waters.

The groundwater would have flowed in and out of the pits, as they were below the Mesolithic groundwater table. This flow of water in and out of the baths was dictated by the moon, and was understood by prehistoric man. No doubt this would have added to their sense of wonderment, and made the site even more magical than it appears today. As a testament to the power of the moon, and to help them predict its movements and, therefore, the daily and seasonal tides, the builders dug 56 holes, which we now call Aubrey Holes in honour of their discoverer. They also sent to the Preseli Mountains in Wales for a supply of bluestone megaliths.

Bluestones – Aubrey Holes

Bluestones were brought by wooden barges the short distance (just 82 miles) from the Preseli Hills in Southern Wales. The higher groundwater levels would have allowed the transportation of the bluestones via a more direct route compared to the current theories that the bluestones were brought by boats sailing around South West Britain, or were dragged overland for hundreds of miles through forest (which would have been impossible, requiring levels of manpower greater than the estimated population of Britain at that time).

Moreover, both these sites would have been on the shorelines of an aquatic forest that covered Britain, and therefore wood for transportation by boat/floater would be plentiful. Prof. Richard Atkinson (Stonehenge , Penguin Books, 1956.) provides an example of how the 7 ton unfinished Altar Stone could be floated on a log boat made of pine with a density in the region of 35 lb/ft^3 (560 kg/m^3). He calculates that a raft of some 700 cubic feet (20 cu. metres) could carry the stone along with a crew of 12 average men. Such boats do not tend to survive the ravages of time, although an example of this type of boat was found in Derbyshire in 1998, which was dated to circa 1300 BCE. It was 11 m (36 ft) long, capable of carrying 4 tons (the weight of a Stonehenge bluestone). If you lashed two or more boats together, you could carry much heavier stones.

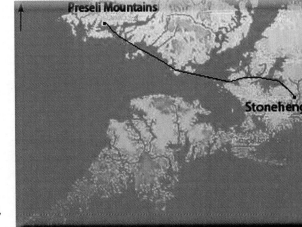

Ice Age Ends   Mesolithic Age Begins

An even easier option is to strap sufficient wood to the stone that it becomes buoyant. The same Altar Stone could be floated with just 10 cubic metres of pine – half as much as the boat – if the wood was simply lashed to the stone. This could then be dragged behind two smaller crewed boats for guidance. The stones, when cut, would be brought down to the shore via a 'log rail' that used levers to move the stones. Once the stones were deposited at the shore at low tide, logs would be attached to the top and bottom of each stone. As the tide came in, the stone would float on its wooden raft. Men could then stand and punt the stone down river, avoiding rocks and sandbanks.

At Stonehenge, the raft would be guided to the shallow North West side of the peninsula, where the groundwater was shallowest and the mooring posts had cross posts attached. The lifting system for getting the stones out of the water would have needed only a minimal number of men, as it used the tide to lift the blocks in the air. The stone was brought to the mooring at high tide and lashed to the cross bar with large ropes. When the tide retreated, the stone would be left hanging above the water, allowing the removal of the flotation logs. A track could then be laid so that the stone could be levered to the site, or a sledge could be brought under the stone to slide it the short distance to the monument.

Either system would require a minimal number of workers: 10 to 20, far less than conventional estimate of 200+ men lugging rocks across Salisbury Plain. This lever system would be second nature to a water-based civilisation. Using trees as their main source of materials, they would quickly have adapted to use wooden levers to manoeuvre their boats, either through punting, or by adding a couple of upright sticks to the side of the boat to create oars that could propel their boats much faster than punting or canoeing.

Palisade and Excarnation

This was the conclusion of Phase I at Stonehenge – a facility that had a dual purpose. Firstly, it was a place of healing; we have seen in previous sections that the stones, when used in water, could heal the sick. But it would have been clear that not all the people who came to Stonehenge could be

timeline

4500 BCE	You are HERE	2500 BCE	800 BCE	0-400 AD	1-2000 AD
Neolithic Age Begins		Bronze Age Begins	Iron Age Begins	Roman Period	Written History

successfully cured. The monument would also have been used as a place for the dead and the journey to the afterlife. Indeed, we still use the same custom and practice today in our modern hospitals. These are our centres for the sick, but hospitals are also the place where we keep the dead in a mortuary prior to their burial. Our Mesolithic ancestors clearly had the same philosophy of sickness and death, and kept them close to each other at Stonehenge.

To understand the construction of Stonehenge, you must be able to interpret our ancestors' beliefs and motives. The dead had to return to the land in the sky, and the only way that could happen was through giving the body to the only creatures that shared the sky with the ancestors – the birds.

Wooden Palisade

Archaeologists have found evidence of these excarnation practices at other sites in Britain. One similarity of these sites lies in the presence of a palisade to protect the bodies from animals other than the birds who fed on the corpses. Such a palisade was found at Stonehenge in the early 1990s, during excavations in the new visitor's entrance. The palisade successfully cuts the natural peninsula off from the heavily wooded mainland to which it's attached.

As there are several main burial mounds on higher ground overlooking Stonehenge, we can only imagine that the entire site was sacred, and may have been completely cleared of woodland so that it appeared much as it does today, laid bare of trees and bushes. One of the remaining 15 barrows overlooking Stonehenge was constructed at the same time, to take the excarnated bones from the site; this area is known as Normanton Long Barrow. Unfortunately, not much of the barrow still exists, but it would have had a distinctive boat shape, surrounded by a moat. This design was a representation of an object and concept the Mesolithic people knew well, as it was their symbol for survival in everyday life.

But the most remarkable aspect of Normanton Downs and Stonehenge is that there are no barrows outside this peninsula, neither below the groundwater line nor West of the palisade. Moreover, even today, the existing tracks reflect the 'ghosts' of the former landscape. If you look at the pathways around Stonehenge, you will notice that they change direction without reason. As most paths follow a straight line between the beginning and the end, we should look for reasons if the track changes course. The path that runs from the car park at Stonehenge to the barrows at Normanton Down tracks over simple grassland; there are no obstacles or field systems to divert the path, yet there are two distinctive changes in direction.

17000 *BCE*   **10000** *BCE*

Ice Age Ends   Mesolithic Age Begins

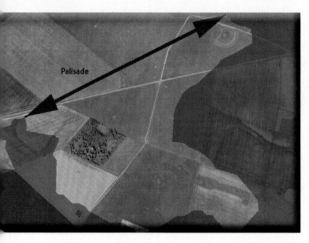

Palisade

When we introduce our predicted shoreline and palisade, the reasons for the changes in path direction become clear. As the path cuts through the palisade, it changes direction and heads for the nearest shoreline inside the peninsula – could this had been one of the old entrances to the site?

Then when the track reaches the groundwater's edge it changes again! This is a flat track in grassland – there is no obvious reason for it not to be straight. If that is not enough, the track turns East and runs along the shoreline – the only reason we see this topology is because the palisade and the shoreline existed in the Mesolithic Period.

Wooden Palisade

Archaeologists know of the palisade and its probable use as a type of shield for the site (in fact, some recently suggested it was a snow barrier!).

But without the groundwater, it doesn't make sense, as you could just walk around the thing to gain access. The only possible reason for the palisade is to join the two areas of groundwater, isolating the peninsula as a sacred place and preventing anything without a boat accessing the site, allowing the dead to go to their maker without being eaten by land animals.

PHASE II Construction – 4000 BCE to 4300 BCE

As the groundwater of Britain slowly started to subside and separate into what would become the Irish Sea, North Sea and English Channel, the landscape of Britain altered significantly. The earlier immense waterways were reduced to lakes and smaller rivers. At Stonehenge, the mooring station at what is now the car park had dried up by the end of the Mesolithic Period, and the site needed to change if it was to survive as it had over the last 3,000 years.

The Avenue

A clear indication of the changes that were introduced in Phase II of Stonehenge can be seen in the development of the Avenue to maintain the ceremonial link between Stonehenge and the river system. Our ancestors started by backfilling the ditch in the North East sector of the site. Both Hawley and Atkinson observed that the secondary filling was not natural. This backfill extended to a depth of 1 m,

timeline

| 4500 BCE | You are HERE | 2500 BCE | 800 BCE | 0-400 AD | 1-2000 AD |
| Neolithic Age Begins | | Bronze Age Begins | Iron Age Begins | Roman Period | Written History |

at which pottery and bluestone fragments are first seen, clearly indicating that it preceded the arrival of the bluestones.

When the Avenue was first constructed, it would have had a dual purpose: to serve as the new mooring processional way for the dead, and also to help maintain the groundwater levels, because the moat had been reduced to a trickle as a consequence of the lower river levels. We have seen in the archaeological section of our hypothesis that Hawley had found a liner in the moat. This liner would not have been required when the moat was first constructed, as the groundwater tables were sufficiently high to fill the moat, whether daily or periodically. But a liner would have been necessary once the tidal groundwater no longer reached the base of the pits.

To keep the moat at a suitable bathing depth, the pits would have needed to be topped up from time to time with groundwater from the receding rivers around Stonehenge. If the prehistoric people took the waters directly from the rivers, it would not have been the clean filtered water that the Stonehenge moat was built for, so they needed to devise a system that let them fill the moat on a daily basis with fresh filtered water. The solution was the Avenue. The Avenue is a processional causeway that had a deep trench built into both sides. This trench is totally unnecessary unless there is another reason for its use, and that reason was to top up the main moat at Stonehenge with filtered water.

Avenue to Neolithic shoreline

The ditch allows clean filtered water to travel along the Avenue all the way up to the existing moat. It would have been simple for Neolithic man to transport this water from the lower Avenue ditch to the higher Stonehenge moat, either by hand or using a wooden hand pump system. It may even be that as a consequence of moving the groundwater so close to the original site moat, it could fill the main moat by natural seepage through the porous chalk rocks – until more engineering studies and excavations of this point are undertaken, we can only theorise. The growth of the Avenue can be seen in the post holes found inside the causeway. There is no good reason to place post holes in the causeway, except to act as a mooring stations to the river. As the river shrank, the mooring stations were moved further away, until eventually they reached the last traditional use of the mooring stations at the 'elbow' of the Avenue.

Sarsen Stones

The Sarsen stones would have been brought to the Stonehenge site before the river disappeared from the immediate plain to become the River Avon of today.   The reason why our ancestors used Sarsen stones is very interesting, as the original structure, which consisted of Welsh bluestones, lasted over 3,000 years as a centre for curing sickness. Clearly, they had a requirement for a very large monument built of a different stone.

From an engineering point of view, the size and structure was very important, and because of the size of these stones, the most effective means of transportation to this site would again be by boat. Even though the deeply forested landscape would have started to thin, the sub-soils would still have been waterlogged and marshy after the groundwater had receded, making it impossible to drag heavy stones across large distances.

The construction of the Sarsen monument is of even greater interest, because if they were building only for aesthetic pleasure then you would imagine that simply laying one giant stone on top of another would be sufficient, as seen at other megalithic sites. But our ancestors wanted to do something special with these stones, so they carved mortise and tenon joints on their surfaces. The only engineering reason for this method of construction would be that the Sarsen stones were to take weight-bearing loads or for the building to last a considerable time, outliving the constructors generation.

The exact date of this process may never be fully known, for we have yet to find massive post holes in the Avenue similar to those in the car park at Stonehenge, which would give us an accurate date. But we still have a few clues to give us an approximate date, as we know that the river must still have been present at the end of the Avenue for the unloading of these stones to take place. And it must have been after the groundwater left the North West shoreline, but before the groundwater reached the 'elbow'. This gives us a date between 6000 BCE and 4000 BCE.

The arranging of the Sarsen stones has left archaeologists without any clues about the dating of Stonehenge. The current theories are based on pottery and the dating of antler picks found in the ditch; all these items could have been left at a later date. For traditional archaeologists to be correct, then the pottery and the antler picks had to be left 'in situ', the archaeologist's way of saying that something has not moved from the place where it was originally deposited – unfortunately, they were not! As an example, the antler picks were found in the ditch. If the ditch had been dug and then left untouched for 5,000 years, then you could have a good accurate date – but we have proved that the ditch was in fact a moat filled with groundwater. These picks would have floated away. All we can say regarding the picks is that the last time the moat was cleaned out was about 2500 BCE – and that's it! And the same can be said about the pottery.

Neolithic Mid-summer Solstice sunrise with stars

Moreover, there is one antler pick that was found under a Sarsen stone! Now this one can't be explained away like the others, as a 12 ton stone was on top of it for thousands of years, guaranteeing that it could not have been placed there at a later date, or floated away to another part of the site. It was found in the 'packing' for Sarsen Stone 27. This gave a carbon dating of 4175 BCE +/- 185, which was consequently dismissed as it did not fall into the same dating range as the antlers left in the ditch, which was supposed to be its contemporary. This date is in the same range as the groundwater table for the Avenue, and therefore should not be ignored.

Let's revisit the Avenue. It clearly follows a path to the river, but the river reached all the way to the East of the site in Neolithic times, and therefore in theory the Avenue could have been built anywhere. So why did they build it in that particular direction? The North West entrance is oriented quite deliberately towards the midwinter moonset, perceived as the place of the dead and afterlife. After 3,000 years, is it possible that the monument changed its purpose? 3,000 years is a colossal amount of time – the same period before now, we were in the Bronze Age living in mud huts and dancing to druid music. In the landscape, we see that the monument's use changed, as round barrows started to appear and burial practices started to change.

Is it possible that Stonehenge as an excarnation site was 'out of date'?

If so, perhaps they decided to still use the trusted waters of the past, and change from curing the sick to celebrating life and the sun? This would explain why the Avenue was oriented to the summer sunrise and, this being the case, give us our third clue to the date the site was built.

We are familiar with the masses that welcome the midsummer sunrise over Stonehenge – people wait in expectation, then (if you're lucky) the sun creeps over the Heel Stone to welcome another day; everyone's happy and goes home drunk. When you look at the Heel Stone, it is on the extreme right-hand side of the Avenue, bent over at a silly angle. Our ancestors did not build it that way – the monument was very nearly completely rebuilt at the end of the last century, and stones were moved.

The most sensible alignment is straight down the middle of the avenue, obviously! This is obvious to anyone with half a brain, so why has it never been questioned or investigated?

The sun does not always rise and set in the same place in history as you may expect. The earth 'wobbles' on its axis in a process known as 'procession' - I will not go into detail here, but all you need to understand is that the Sun and moon rises and sets in a different location over a period of 43,000 years.

This means that the summer and winter Solstice moves in the relation to the horizon a fraction every year. This movement is TINY. Its 0.0002 of a degree every year, but over a long period of time, say 10,000 years ago it's a whole two degrees. It may not sound much, but when you consider that the moon is half a degree in diameter, then two degrees is the same as four moons (or suns) in a row on the horizon.

We can also 'reverse engineer' this figure to give us a date for the construction of the Avenue. The centre of the Avenue is 49.21º ; the current Summer Solstice Sunrise above the Heel Stone is 50.62º - this is a difference of 1.41º. If each year the sun moves 0.0002247 (exactly) then we will have the exact date (when the sun is at the centre of the Avenue) and construction date of 4275 BCE.

Station Stones

The Station Stones seem to have been added during Phase II of the site's development. Aubrey post holes would have been obscured by the introduction of the Station Stones; therefore, they must post-date the original bluestone circle but predate the filled moat, as they have moats of their own. Whether the knowledge of the tides was no longer necessary, or the station posts held a special purpose, we currently don't know. Modern theories use astro-archaeological alignments based on these posts to speculate on the

timeline

4500 BCE	2500 BCE	800 BCE	0-400 AD	1-2000 AD
Neolithic Age Begins	Bronze Age Begins	Iron Age Begins	Roman Period	Written History

reasons for their existence. Unfortunately, only two of the four Station Stones have a moat, which really does not make sense if they are as important as these theorists believe. These moats were directly connected to the main moat, as we have proved in the previous section. Our ancestors built barrows as signposts, not burial mounds. These markers are aligned to show not only where to go but, more importantly, how to get home.

Does it also follow that they would use the same method to point the way to the afterlife?

If you lived in the countryside or became a fell-walker in the days before GPS and OS maps, you used to have to rely on points on the horizon for guidance. The same simple principle would have been used by prehistoric man to get from A to B without getting lost. Initially, these features would have been on islands, as people used boats to transport themselves and trade. Then, as the groundwater fell, they would have used barrows as markers on the horizon to walk from point to point. We still see milestones today on the side of roads; barrows were the milestones of prehistoric man.

So, looking at the Station Stones, where would that direction have taken our ancestors?

Well, if you follow the line from the centre of the site through the Northern Station Stone, you will go past no less than 5 long barrows, 15 round barrows, Casterley Camp, Knap Hill Camp, and the White Horse, finally arriving at Avebury. This is not bad, for just 36.4 km of travel. That's one barrow every 500 metres; not even I could get lost with that frequency! In fact, mathematically, the chance of this number of barrows being in line over such a small distance is less than half of one per cent (0.05 %), or 2000 to 1 in layman's terms.

The Southern Station Stone points the way to Old Sarum near Salisbury. Although it is famous for being the site of the original Salisbury Cathedral, archaeologists have also found evidence here of flint tools that date back to 3000 BCE. We believe that there is a clear indication that Old Sarum was first used in the Mesolithic Period, when it was an island above the raised groundwater line, and that (as in many cases) later sites were built upon the site of this original construction. We look in-depth at this site later in this section.

The most interesting of all the original markers at Stonehenge must be the Heel Stone. The Heel Stone is slightly right of centre in the Avenue and, like two of the Station Stones, it has its own moat. If we line ourselves up with the Heel Stone from the centre of Stonehenge, it aligns with Durrington Walls and Woodhenge. Clearly, these three places would have been not only the most important neighbouring sites to Stonehenge, but a gateway to other sites and trading places in the ancient world. Next to each of the moated Station Stones, there was a gap left in the moat to allow people to cross by the stone in the direction it indicated.

Use the same method to point to the afterlife?

The station stones - which direction do they point?

You are HERE

17000 BCE

10000 BCE

Ice Age Ends

Mesolithic Age Begins

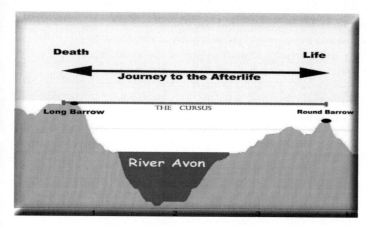

This no doubt led the walker along a path, now lost to us 5,000 years later, via a system of barrows, to the desired location.

The Cursus

If you draw a line from the Preseli mountain region in South Wales to Stonehenge, you will find that your line not only points to the horizon where the midsummer moon rises, but also passes a dozen ancient long barrows on the way. It leads over the far Western edge of the Cursus, past a point believed to contain yet another crescent Bluestone circle facing North West, which was first suspected by Atkinson but which has yet to be located. Now, if you do the same exercise with the direction of the summer solstice sunrise, drawing a line from the centre of Stonehenge through the Heel Stone, you will see on the horizon another long barrow which, though it has now been ploughed out, marks the other end of the Cursus.

So we have an alignment from the centre of Stonehenge to the Preseli Mountains that cuts over the Western end of the Cursus, and an alignment over the Eastern end of the Cursus to Woodhenge. If you examine the profile of this area, the Cursus cut across a valley that we have now proved was full of groundwater in the Mesolithic Period. This would have given the monument the appearance of two islands separated by a RIVER!

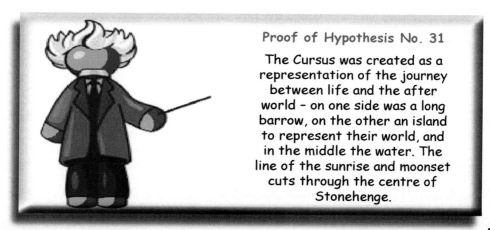

Proof of Hypothesis No. 31

The Cursus was created as a representation of the journey between life and the after world – on one side was a long barrow, on the other an island to represent their world, and in the middle the water. The line of the sunrise and moonset cuts through the centre of Stonehenge.

timeline

4500 BCE	2500 BCE	800 BCE	0-400 AD	1-2000 AD
Neolithic Age Begins	Bronze Age Begins	Iron Age Begins	Roman Period	Written History

This being the case, it does not take much imagination to see that when the groundwater receded during the Neolithic Period, to maintain the water connection between the two end sites of the Cursus, they dug a ditch that encapsulated the entire region. What we see in the Cursus is a giant model of the world as our ancestors perceived it. On the Eastern side was a long barrow representing death, where the bones were stored. In between lay the water and the final journey to the afterlife. On the Western bank of the Cursus was an island, probably marked with a large standing stone, where the ancestors arrived after death, no doubt to be reborn again. What we see at the Cursus is the first sculptural artwork in the history of mankind, and the size and magnitude of this artwork gives us a clear indication of the resourcefulness and dominance of this great civilisation.

Palisade

We should not forget that Stonehenge is incredibly old, and some of its features would have been changed to adapt to the new landscape. Therefore, when the groundwater table dropped at the area of the site that is now the car park, the palisade would have failed in its function as a barrier against animals. There is clear evidence that, the palisade was extended to the elbow of the Avenue. The evidence indicates that although the number of excarnations at the site began to fall, the desire to separate the peninsula from the mainland may have continued.

PHASE III Construction – 3500 BCE to 2500 BCE

At the end of the Neolithic Period, the groundwater had subsided to almost their present levels. The large river that had existed at Stonehenge for 5,000 years, since the great Ice Age melt, was gone. The groundwater that once covered the land had moved to the surrounding seas and channels, flooding the island once called Doggerland and leaving it 30 metres under the sea. At Stonehenge, the moat had dried up and was dug out for the last time for ceremonial purposes. The tools used were left where they broke in the ditch. Archaeologists now take these tools as the basis for their incorrect dating of the site, not to when the monument was at the height of its power, but to the time when it was last used as a sacred site.

The Avenue

Once the groundwater had gone from Stonehenge Bottom, and the Avenue had become a path to nowhere, the builders clearly decided it was time for a complete makeover. They added to the original walkway a path to the River Avon, which would then have been only a few metres higher than today. The path almost turns back on itself, showing clearly that it was not all constructed at the same time, otherwise the builders would have taken a more direct route. Why they bent the path around to the South, rather than carrying on Eastwards to the River Avon much nearer Woodhenge, is again unknown and we can only speculate. If their ceremonies were still connected to boat travel, it would take a longer trip to and from Woodhenge to fulfil their need. Perhaps they had found a site for their bluestones which had a direct source of clean water.

Bluehenge

Professor Mike Parker-Pearson found, in 2008, the outline of a stone circle by the banks of the Avon next to the extended Avenue causeway. It was reported in the Daily Mail that:

"The monument has been tentatively dated to between about 3000 and 2400 BCE."

Excavation revealed several stone settings that are thought to have been erected around 3000 BCE. It is estimated that there may have been as many as 27 stones in a circle 33 feet (10 m) wide. The name "Bluestone henge" is derived from the discovery of small stone chips in some of the stone settings. These bluestones are also found in Stonehenge and consist of a wide range of rock types originally from Pembrokeshire West Wales, some 150 miles (240 km) away. Archaeologists suspect that bluestones in the circle may have been removed around 2500 BCE and incorporated into Stonehenge, which underwent major rebuilding work at about this time.

The stone circle settings were surrounded by a henge, comprising an 82 feet (25 m) wide ditch and outer bank which appear to date from approximately 2400 BCE. Unlike Stonehenge, there do not appear to be any significant solar or lunar orientations within the monument." There are two very relevant and interesting points about this find. Firstly, it proves our hypothesis about the use of the bluestones with water, and consequently the necessity of locating them next to the Avon. Moreover, using our hypothesis, we can date this sequence more accurately, as we know that at the time of the construction of Phase I and II of Stonehenge, the Bluehenge location was under about 30 metres of groundwater. The Bluehenge circle must, therefore, be later than Stonehenge in date. What Bluehenge clearly indicates is that the Stonehenge area continued to be used as a centre for healing the sick even into the Iron Age, but provided its healing treatments at an associated location connected by the Avenue.

Round Barrow Alignments

As we have already stated, the archaeologist Alfred Watkins noticed that some ancient sites and barrows were in alignment with one another. We have noticed that Stonehenge's three moated stones are aligned to three of the most important sites in the area: Avebury, Old Sarum, and Durrington Walls and Woodhenge. Further investigation has led us to believe that more alignments occur with round barrows over land. This would have been the only sensible way for prehistoric man to navigate over land, as maps had yet to be discovered, and we know that all ancient civilisations used points on the horizon as reference points.

This would also explain why not all barrows contain skeletal remains.

4500 BCE	2500 BCE	800 BCE	0-400 AD	timeline 1-2000 AD
Neolithic Age Begins	Bronze Age Begins	Iron Age Begins	Roman Period	Written History

PHASE IV Construction – 2500 BCE to 1500 BCE

Bluehenge

Even after moving the bluestones down the valley to the Avon, it seems that Bluehenge was abandoned not long after its construction. Perhaps the waters of the Avon flooded the location, or the knowledge of how the stones cured the sick was lost or superseded. It is likely that the stones were either returned to Stonehenge, or broken up and used for building during the Roman to Medieval period.

Heel Stone Alignment

During the later part of Stonehenge's prehistory, it lost its initial function as a mortuary, as our ancestors beliefs changed and burial practices altered from excarnation to burial. They still, however, needed a monument to reflect the voyage to the afterlife. At some time in the Iron Age, or after the groundwater had dried up, the Heel Stone that once indicated the passage to the old rebirth site at Woodhenge was moved to align with the sunrise. It was deliberately tilted slightly over the course of time, to match the changes in sunrise direction caused by the very slow movement of the Earth's axis, so that the midsummer sun rose from behind the stone.

Chapter 14 - Old Sarum - the missing link (Case Study no.2)

If we use the Southern Station Stone as a direction marker from Stonehenge, it points South by South East. If we then follow the line on a map of our Mesolithic landscape, showing the raised groundwater tables, we find that it points to an amazing site just 10 km away from Stonehenge: an island in the middle of a very large waterway. This island would be found quite easily by boat. But even so, we have found that barrows and sites serve as markers on the shoreline, to allow boats to navigate to this island from Stonehenge even at night or in bad weather.

The positioning of this site is most interesting, as it lies very close to the Channel which would have taken boats off to the continent. It is quite possible that this was Britain's first sea port, as it seems to be the last known occupied Mesolithic island. If so, it would have been one of the most important sites not only in Britain but in what we call Europe, as it would have been involved in nearly all Mesolithic imports and exports to and from France, Spain and the Mediterranean countries.

The island of Sarum would have been a magnificent site, sitting in a huge river as wide as the eye could see. At the time of Phase I of Stonehenge's construction, the groundwater around Sarum would have reached higher than the outer ditches we see at Old Sarum today. This gives us our first clue as to when the ditches were constructed, as it would be an impossible task to dig ditches if the area flooded twice daily. It also explains one of the anomalies at Old Sarum, a large deep ditch in the South of the site that serves no defensive purpose and was originally believed to be an old landslide.

Harbour A

Harbour B

Mesolithic Old Sarum showing it was an island

When our Mesolithic groundwater is introduced, this feature is flooded and becomes an ideal mooring station for boats. The hollow is huge, and could have harboured 40 to 50 small boats or a couple of 20 metre ships! There seems to be another similar, somewhat smaller, feature to the North West of the island. The lack of detailed excavations of this site prevents us from offering further archaeological proof of our hypothesis. Current theories on this site speculate that the central moat was dug for the motte-and-bailey castle that stands there today, but without firm evidence this is just guesswork.

timeline

4500 BCE	2500 BCE	800 BCE	0-400 AD	1-2000 AD
Neolithic Age Begins	Bronze Age Begins	Iron Age Begins	Roman Period	Written History

Proof of Hypothesis No. 32

The depressions to the North and South of Old Sarum show that, in Mesolithic times, these features were used as harbours on the island that was Old Sarum.

Looking at this island in Mesolithic times, the raised groundwater tables would have filled the original inner ditch, which is some 7 metres deep, in the same way as the moat that surrounded Stonehenge in Phase I of its construction. We therefore suggest that the moat was first dug in the Mesolithic Period. The Romans, and then the Normans, would have used the existing moat or enhanced it for their own requirements. The current archaeological belief that the Romans or the Normans dug the ditch does not stand up to scrutiny, as the outer ditch already existed at the site during their occupation, and it would have been simpler and cheaper for them to use the existing structure than to ignore it and make another.

The site at Old Sarum is much bigger than Stonehenge, and is of a similar size to Avebury. One can only guess at what would have been in the centre of the Mesolithic island. From the amount of reused Sarsen stone found in the remains of the Norman castle and the original cathedral, we can infer that a megalithic structure like Stonehenge or Avebury stood at Sarum during prehistoric times. If you extend a line from the centre of the motte-and-bailey at Old Sarum through the centre of the church (the original Salisbury Cathedral), it points to Stonehenge.

Churches built on prehistoric sites are not uncommon. There are many instances of pagan religions crushed by Christianity taking over their sacred sites and using the stone circles as building material for their churches. So, we believe that in prehistory three stone circles existed at Sarum: one large outer circle and two smaller inner circles, indicating the way to Stonehenge. Later, as the groundwater fell, our ancestors built Sarum's outer banks to keep their sacred site an island.

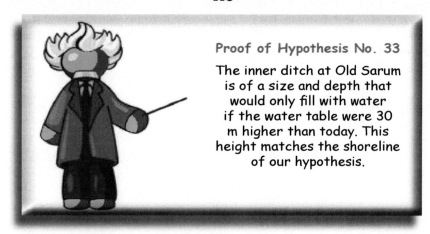

Proof of Hypothesis No. 33

The inner ditch at Old Sarum is of a size and depth that would only fill with water if the water table were 30 m higher than today. This height matches the shoreline of our hypothesis.

At Avebury, a similar configuration can be seen, with two smaller stone rings inside the large ring that borders the outer moat.

In the Neolithic Period, the groundwater table dropped by about 10 m, and the island of Sarum was joined to the mainland by a peninsula. Our ancestors therefore built giant ditches 12 m deep, to keep the site surrounded by groundwater. The Southern and Northern mooring-points could no longer be used, as the groundwater had receded too far, so the Neolithic people created a new landing point to the West. They left a gap in the huge ditch, so that people and goods could enter the island; this would have looked very much like a bridge across the water.

At the end of this footpath, they built another mooring station that protruded from the edge of the moat like a peninsula, so that boats could be moored safely around the feature. For some bizarre reason known only to early archaeologists, the platform is shown on some maps as a Roman road connected to a road some 200 metres away on the West side of the island. Unfortunately, for that theory to be correct would take a leap of faith and nature. The landing platform, which is shown as a lumpy protrusion on maps, has a 1:2 slope, with a vertical drop of over 30 metres. I would suggest that a Roman horse and cart would not be an advisable means of transport for this terrain unless they were equipped with ABS brakes and a parachute!

The historical record does give us some clues to Old Sarum's deeper past and the ways in which the groundwater that surrounded the island dictated its history. The original Salisbury Cathedral was built here, only to be moved down to the valley a few hundred years later. Can you guess the reason for the move? That's right, the lack of water! It seems that even over the cathedral's brief history at Old Sarum, the groundwater continued to subside. As this story is well known, why did no-one wonder how deep the rivers might have been thousands of years ago?

timeline

4500 BCE	*2500* BCE	*800* BCE	*0-400* AD	*1-2000* AD
Neolithic Age Begins	Bronze Age Begins	Iron Age Begins	Roman Period	Written History

The Maths

Currently, the groundwater table around Old Sarum is 56.5 m above sea level. The well in the Norman fort is 70 m deep from an altitude of 130 m, which shows that the groundwater is today 3.5 m below the Norman well. Therefore, the groundwater table in 1000 AD - when the well was first constructed - must have been at least 60 m, so in 1,000 years the groundwater has fallen 3.5 m. If we multiply this thousand-year drop in groundwater table by 9, then add 56.5 m to account for the existing groundwater table, we can estimate the groundwater table 9,000 years ago, i.e. in 7000 BCE. That would make the groundwater table (9 x 3.5 m) + 56.5 m = 88 m.

The outer banks of Old Sarum are 89 m above sea level – close enough, I think!

Proof of Hypothesis No. 34

The history of Old Sarum shows that the water table dictates the uses and functions of the prehistoric island. If we look at the receding water levels at Old Sarum when Salisbury Cathedral was abandoned, we can reverse engineer the water levels during the Mesolithic Period – they match our hypothesis.

Chapter 15 - Avebury - the oldest of them all (Case Study No.3)

The Northern Station Stone of Stonehenge is in direct alignment between the centre of the Circle and Avebury.

Avebury lies in an area of chalkland in the Upper Kennet Valley, at the Western end of the Berkshire Downs, which forms the catchment for the River Kennet and supports local springs and seasonal watercourses. The monument stands slightly above the local landscape, sitting on a low chalk ridge 160 m (520 ft) above sea level; to the East are the Marlborough Downs, an area of lowland hills. Archaeologists freely admit that the history of Avebury before the construction of the henge is uncertain, because little datable evidence has emerged from modern excavations. But stray finds of flints at Avebury, dated between 7000 and 4000 BCE, indicate that the site was visited in the late Mesolithic Period.

If we now apply the same groundwater table adjustments demonstrated in our Stonehenge case study, we are left with a landscape rendered unrecognisable by groundwater, as the Avebury Circle becomes an island. The most remarkable thing is that Avebury now looks like a sister site to Old Sarum: both are perfectly round islands surrounded by groundwater; both have two inner circles and are aligned to Stonehenge via its moated Station Stones.

The next item of interest is the orientation of the long barrows. If you look at an OS map of Avebury, you can see that the barrows are not all oriented in the same direction. Archaeologists would have you believe that these monuments were only made for the dead but, if that was the case, why don't they point to a particular direction, such as the sunrise or sunset, or something equally symbolic? From our Mesolithic groundwater map, we can show that East Kennet Long Barrow was the first hill marker you would see if approaching Avebury from the Eastern inlet. Although West Kennet Long Barrow is seen side-on, it would still be visible as a smaller marker, as it had large white stones added to its Eastern entrance to give it greater visibility.

When the groundwater started to recede, as we saw at Old Sarum, our ancestors tried to keep their monument an island by adding ditches. These ditches would have been shallow at first, becoming deeper over the centuries until they were finally abandoned, leaving what we see today.

This gradual process explains more clearly how and why such a task was undertaken, as the logistical requirements of building the Avebury ditches 'in one go' would have been beyond a prehistoric civilisation whose only tools were antler picks and stone axes.

Current estimates suggest that it took 1.5 million man hours to build the Avebury monument. In simple terms, that's 200 people working full-time for 3 to 4 years. This is clearly not plausible. As you will be able to imagine if you have ever visited the site or you understand the requirements of manual labour, it would take a lot more to construct such a large site with such basic tools.

The nearby man-made Silbury Hill contains 248,000 cubic metres of chalk, and would have taken 18 million man hours to construct (Atkinson 1974:128). That's equivalent to 500 people working full-time for 15

timeline

4500 BCE	2500 BCE	800 BCE	0-400 AD	1-2000 AD
Neolithic Age Begins	Bronze Age Begins	Iron Age Begins	Roman Period	Written History

Proof of Hypothesis No. 35

The Long Barrows around Avebury are built on the shorelines of the Mesolithic waterways predicted by my hypothesis. Not only are they on the shoreline, but their orientation proves that they were used to navigate ships between Avebury and Stonehenge.

years. Yet we are expected to believe that Avebury's 125,000 cubic metres of chalk took just 1.5 million man hours to move.

There is just no consistency in archaeological findings; it's all subjective, and quite frankly wrong!

It's more likely that these monuments grew over the course of centuries, slowly but surely, the ditches starting at just one metre but getting deeper over the next 5,000 years as the moat was cleaned out, until they reached their final dimensions of 11 metres deep and 22 metres wide. This gradual process would explain another archaeological mystery that the 'experts' avoid - with what tools were they built?

Now everyone knows that Stonehenge, Avebury and Old Sarum were cut out of the hard chalk with antler picks - or do we?

For if the archaeologists are right, the entire site, must be littered with the broken remains of these objects - but they're not!

Half of Stonehenge has been fully excavated and found just 82 pieces probably from about 50 full antlers. At Avebury even less - either antlers are the hardest natural tools in the world or what we find are the remnants of tools used after the construction for alterations or to clean out (not cut) the ditches.

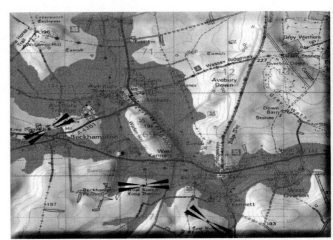

Avebury's three Long Barrows and orientation

Proof of Hypothesis No. 36

The ditches at Avebury are over 11m deep. The excavated chalk was not used to bank the inner side of the ditch as expected if it was a defensive feature – but on the outside to shelter the circle. Therefore, the only possible reason for such an excavation would be to create a moat for boats to harbour.

So if antler picks didn't dig the ditches, what did?

Mike Parker-Pearson found strange cut marks in the bottom of a ditch at Durrington in 2008, the cut marks were so thin that they could only be made by a metal blade, like an axe, the only problem is that according to traditional archaeology, this Bronze technology would not be available during this period of the stone age - unless, of course the accepted dating periods of our history are fundamentally wrong and metals were invented long ago and the techniques re-invented in the 'Bronze Age'. For we now know that the peoples of Europe had Bronze as early as 4600 BCE in Bulgaria found in a Gold and Bronze grave - so there is no good reason to believe that the builders of Stonehenge did in fact have Bronze axes to cut the chalk.

Avebury with high bank and water to protect boats

Back at Avebury, the most interesting aspect of this excavated 'chalk mountain' is that we can use it to approximate a date for its construction and therefore the excavation of the ditch the chalk where it originated.

We know the original height of chalk bank as we know how much chalk was removed from the ditch in front of the mound.

So if antler picks didn't dig the ditches - what did?

timeline

4500 BCE	2500 BCE	800 BCE	0-400 AD	1-2000 AD
Neolithic Age Begins	Bronze Age Begins	Iron Age Begins	Roman Period	Written History

Proof of Hypothesis No. 37

The erosion of the chalk banks at Avebury give us an indication the approximate date the monument's ditches were first constructed. At 1mm per annum this indicates the ditches were built BEFORE 6000 BCE.

On average the ditch is 11m deep and 22m wide - which means that for every metre excavated 132m² was transferred to the bank. The chalk mound is much narrower than the ditch from which it was excavated. Therefore, the bank would be higher than the ditches depth - my calculation shows that the bank would be

therefore 15m high at least but as the chalk in the ground had been compressed over time and the bank would have contain more air gaps the likelihood is that the bank was 20m high at the time of completion.

For our calculation, if we take the smaller value of 15m then we see that there is only 7m left today (on average on the banks that still exist or have not been cut in recent times). We know from monitoring the erosion levels of chalk soils to be a nett loss of 1mm per annum (let loss takes into consideration both water erosion of the chalk and soil deposits dropped through rain) **This simple calculation gives us the answer of the bank being built AT LEAST 8,000 years ago - about 6000 BCE.**

Eventually, when Avebury lost all of its groundwater, our ancestors built Silbury Hill as the new landing site to the complex. Silbury Hill is the largest man-made island in Europe,

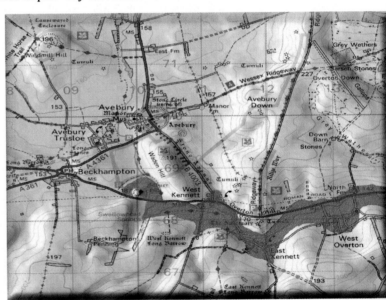

Neolithic Avebury and Silbury Hill

17000 BCE	10000 BCE
Ice Age Ends	Mesolithic Age Begins

and was set at the end of the Neolithic waterway. Composed mainly of chalk and clay excavated from the surrounding area, the mound stands 40 metres (130 ft) high and covers about 5 acres. As we have already seen, it would have taken 18 million man hours to deposit and shape this vast pile of chalk and earth on top of the natural hill that forms Silbury's foundation. The base of the hill is circular, 167 metres (548 ft) in diameter. The summit is flat-topped and 30m (98ft) in diameter.

At this point, to maintain the link to Avebury, the Sanctuary was created and a stone causeway was introduced from the landing site at the end of the peninsula to Avebury. It should be noted that the original track went North East from the Sanctuary. This path would have taken our ancestors around the groundwater and over the hill to a place called Falkner's Circle, which has now been destroyed, but once overlooked Avebury. From this stone circle, the path went down to the existing stone avenue toward Avebury. This stone avenue, like the one at Stonehenge, has a strange kink in its design, as if it went in a different direction before being changed at a later date. The kink shows that the original path led to Falkner's Circle, and we believe this was the route our ancestors used in the Neolithic Period.

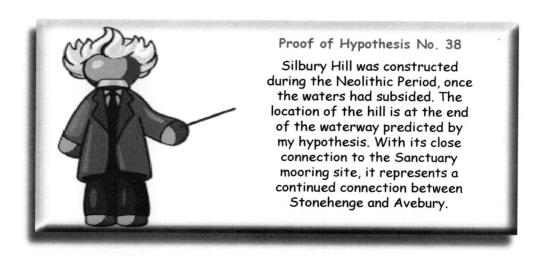

Proof of Hypothesis No. 38

Silbury Hill was constructed during the Neolithic Period, once the waters had subsided. The location of the hill is at the end of the waterway predicted by my hypothesis. With its close connection to the Sanctuary mooring site, it represents a continued connection between Stonehenge and Avebury.

timeline

4500 BCE	You are HERE	2500 BCE	800 BCE	0-400 AD	1-2000 AD
Neolithic Age Begins		Bronze Age Begins	Iron Age Begins	Roman Period	Written History

Chapter 16 - Woodhenge and Durrington Walls a structure by a harbour (Case Study No.4)

If you go to the original North East path break in the Stonehenge ditch and look to the Heel Stone, which was moated during the Mesolithic Period, you will have a direct alignment to Stonehenge's nearest neighbours, Woodhenge and Durrington Walls. The fact that the Avenue is wider than the break in the ditch around Stonehenge has always been a mystery to archaeologists. This mystery is compounded by the Heel Stone's position to the right of the Avenue, rather than in the centre. As shown in our Stonehenge case study, the Avenue was built after the Mesolithic ditch was created; therefore, these individual moated stones start to make some sense as alignment points. The Heel Stone was clearly positioned prior to the Avenue's construction; the moated stone points to Stonehenge's sister site Woodhenge and not, as is currently believed, to the midsummer sunrise.

The first thing that strikes you when you look at Durrington Walls is that it seems incomplete; it looks like a half-circle from aerial photographs, and from the ground you get a sense of it only being half finished.

But most illustrations include the Eastern section, because magnetometer surveys show that under the surface there are more ditches although you might question their purpose, as it is not obvious. The Eastern side of the site was clearly built much later than the original West side. The East bank is smaller and does not match the specifications of the original ditch and moat, which was roughly 5.5 m deep, 7 m wide at its bottom, and 18 m wide at the top.

The bank was 30 m wide in some areas. The bank and ditch indicated by the magnetometer surveys

Legend

- Henge: Bank
- Henge: Ditch
- Henge: Former Entrance?
- 2006 Trench (inc. access ramps)
- 2005 Trench
- 2004 Trench
- 2006 Exc. Geophysical Trench
- Northern circle
- Southern circle

0 20 40 60 80 100 m

Durrington Walls

Woodhenge

Durrington Walls before the extension in the Neolithic Period

Durrington Walls in the Mesolithic Period

17000 BCE   10000 BCE   **You are HERE**

Ice Age Ends   Mesolithic Age Begins

are less than half that depth; the bank is only about a third of the size of that on the Northern side. The current theory and plan of Durrington Walls simply does not stand up to investigation, for it is clear that the Eastern side of the camp was added at a later date, when the prehistoric groundwater had started to recede. This would include the Southern circle found in the 1960s.

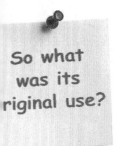

So what was its riginal use?

So what was its original use?

To answer that question, you must look at the site's terrain, position and layout. The first thing that hits you is that the site is not flat! In fact, it's a huge bowl. Archaeologists say that it is a settlement, but anyone who goes camping will tell you not to pitch your tent on a slope, and for a very good reason: you will wake up one morning covered in water, as when it rains the water runs downhill! So, were Mesolithic people completely stupid? If this was an encampment, the houses would flood frequently.

Archaeologists will insist that, because they have found the foundations of a couple of roundhouses, the site must unquestionably be a settlement – this is because they are dumbfounded by its position and shape. If we now add the higher Mesolithic groundwater tables as discussed in our other case studies, the site becomes a perfect natural harbour, with shallow sides for pulling boats ashore and a 4 metre deep ravine in the centre of the harbour that would have provided shelter from gales.

Woodhenge has two entrances: one clearly directed towards Durrington Walls camp and, more importantly, a mysterious second entrance that trails to our Mesolithic shoreline. This is a clear indication that groundwater was present at Woodhenge during Mesolithic times. Not only would it explain the strange shape of the camp, but also the magnetometer survey showing a ditch dug after the groundwater had fallen in Neolithic times. Even more interesting is how the landscape reflected the receding shoreline over 5,000 years at this site. The present-day minor road runs along the course of the Mesolithic shoreline circa 6000 BCE.

We should not be too surprised by this, as lake shores and coastlines still have paths along them today so that we can fully enjoy them. There is no reason to believe that prehistoric man did any different 8,000 years ago, and such a path would also have a practical purpose as the shorelines were used as a mooring site. If we are correct about the road and the mooring points, is it possible to find the same type of post holes here as we found in the car park at Stonehenge?

South to North Cross section of Durrington Walls

timeline

4500 BCE	2500 BCE	800 BCE	0-400 AD	1-2000 AD
Neolithic Age Begins	Bronze Age Begins	Iron Age Begins	Roman Period	Written History

Proof of Hypothesis No. 39

Durrington Walls shows clearly that it was originally constructed not as an encampment, but as a harbour. The V-shaped floor and sloping profile is perfect for mooring ships and boats, and the high banks on three sides would give shelter. This also proves that the waterline in the Mesolithic Period was at the height proposed by my hypothesis.

Unbelievably - Yes it is!

Unbelievably, yes, it is!

Wainwright, in his excavations of Durrington Walls, discovered lots of them. Without the shoreline, which we have added in the diagram, they really do not make much sense. But as soon as the groundwater is added, their function becomes obvious: they held mooring posts, for that is the natural landing area for boats coming to and from Stonehenge.

Receding Shoreline over next 5000 years

Post holes found at the shoreline of Durrington Walls

17000 BCE   **10000** BCE   You are HERE

Ice Age Ends   Mesolithic Age Begins

Conclusion

In this last section, we look at the civilisation that built these monuments that have survived 10,000 years, and we try to understand their culture and technology. In doing so, we are connecting with our ancestors and, probably for the first time, attempting to understand their ways and methods.

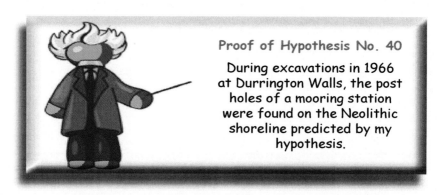

Proof of Hypothesis No. 40

During excavations in 1966 at Durrington Walls, the post holes of a mooring station were found on the Neolithic shoreline predicted by my hypothesis.

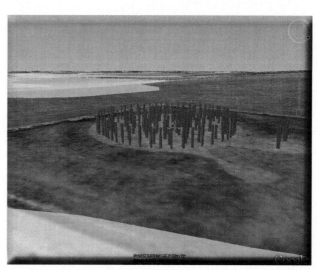

Woodhenge in the Mesolithic Period - showing the shorelines in white

timeline

4500 BCE	2500 BCE	800 BCE	0-400 AD	1-2000 AD
Neolithic Age Begins	Bronze Age Begins	Iron Age Begins	Roman Period	Written History

Section Four: An Ancient Civilisation

In this section we will try to illustrate what kind of culture could organise and build such enormous monuments that lasted ten thousand years. We have shown that these monuments were not constructed 'overnight' by hundreds of workers slaving without rest to reach their endeavours, but by a more controlled, gradual construction process over many hundreds or even thousands of years as their culture and society grew.

The way to measure the greatness of our prehistoric ancestors is through the longevity of their civilisation, for time shows that they had a stable and structured society that kept faith with the same ancient traditions as their forefathers for many thousands of years, like the Native North Americans Indians before our technological civilisation invaded their land. Consequently, this may be our second measure to the differences in our cultures, for today we seek to advance through technology, as we believe the better the technology, the better and greater the civilisation.

But is that really true?

Archaeologists date history by technological phases and not by cultural values; Stone Age, Bronze Age and Iron Age are the main prehistoric evolutionary time periods. This type of technology-focused dating implies that Stone people were more primitive than the Iron Age people because they had LESS lethal metal weapons. This is deeply misleading. Today, our governments can destroy whole continents just by pressing a switch to launch rockets that can kill millions of innocent faceless people.

Is this really progress and a sign of a great civilisation?

Isn't there more honour and a greater 'morality', as a human being, when looking into your enemies' eyes and, in that brief moment, judging whether they are guilty of a crime you deem so serious as to justify taking their life? So, when we evaluate historic civilisations, we must also evaluate our own values and judgements. For if we do not we may fail to understand the true nature of these cultures of the past.

Chapter 17 - The Megalithic Builders

Open any book about prehistory that has been published in the last 100 years, and you will get a 'stereotypical view' of fur-clad cavemen who lived before and after the last ice age 12,000 years ago, chipping away with flint stone tools, eating raw meat and each other. This primitive basic lifestyle supposedly lasted until the Bronze Age in about 2500 BCE, when they cut down all the trees and built strange monuments to appease and worship the gods. The British tribes were then gradually civilised and re-educated by the influences of trade, and later by more advanced invading civilisations that had developed in mainland Europe from the influences from the Middle East. This (in a nut shell) is the current 'Out of Africa Theory', which eventually arrives in Britain and a civilisation in the form of farming and living in mud huts with their pigs and goats.

And is complete nonsense!

This 'dribble' has been regurgitated so many times in the past that we have almost taken it as truth. Consequently, when an archaeologist finds something that doesn't fit this 'norm', it is rejected totally out of hand. In previous sections, we have seen the way good scientific evidence is dismissed when it doesn't fit the old accepted theories of British history. The same disregard is given to our ancestral culture. Academics paint them as unsophisticated fur-clad nomads who hunted wild animals while their women gathered berries; hence the term 'hunter-gatherer'. But then, suddenly, out of the blue, this strange group of farmers, who were incapable of building a decent wooden cabin, got together to build a stone monument, using woodworking techniques, that would last 10,000 years.

We should not take lightly their ability to build something of such magnitude as Stonehenge, as the degree of organisation required to produce these constructions is very rarely seen in our history. There is a tendency for archaeologists to look at certain modern cultures and compare these people with our ancestors. The classic example is the hunter-gatherer tribes of Africa, America or Asia. This analogy is fundamentally flawed, as these tribes do not possess the organisation or the engineering ability to build monuments as large, or as long-lasting, as we see in Northern Europe.

So, how do we establish a framework for us to understand this unknown civilisation?

So how do we understand this unknown civilisation?

The answer is to look even more closely with an unprejudiced mind at their technology, science and mathematics and, through this window in our imagination, we will try to find some answers to our questions. We have already established that this civilisation existed for thousands of years (much longer than our very own civilisation that dates back to the farming revolution just five thousand years ago), continually working on their monuments to adapt them to the falling groundwater of the Neolithic Period which ended their aquatic way of life.

timeline				
4500 BCE	**2500** BCE	**800** BCE	**0-400** AD	**1-2000** AD
Neolithic Age Begins	Bronze Age Begins	Iron Age Begins	Roman Period	Written History

We understand from the man hours they put into building their monuments that they had to be highly organised, as they only had small numbers of people to work at raising stones and digging moats

So, did they all live in caves as portrayed?

Well, if they did, there must have been a lot more caves back then than now, or they would have been rather crowded! Clearly, this is not the answer. Our hypothesis explains that after the last ice age, the land was flooded with groundwater from the melting ice caps. Britain then became a nation of islands with a mild climate, much warmer than today - the kind of subtropical environment now seen in some regions of the South America. After the last ice age, as the tundra that restricted growth receded, trees grew in abundance and the dry land became covered with woods and thick forests. If we are going to investigate the lifestyles of Mesolithic and Neolithic man, we must look to the boat people of the Amazon and Far East for examples of their ways.

So did they live in caves?

Archaeologists have reconstructed roundhouses made from mud and straw, which they suggest is the logical construction type based on the discovery of post holes at some Bronze and Iron Age sites.   The problem is that this type of house seems to have come into being in the Bronze Age, with no evidence of earlier man-made housing in the Mesolithic and Neolithic; hence the association of cave dwellings with this period, which spans about 7,000 years (5 times longer than the 1,500 years of the mud hut phase). This lack of accommodation is even more bewildering when compared to the elaborate housing found in the forsaken wasteland of the Orkneys. The round stone structures found there, which had built-in furniture and walls 2 metres high to keep the elements out, are of a greater sophistication than any Bronze and Iron Age mud huts in the South, but yet these houses are older.

So, if we had houses in the past, why did we go back to mud huts?

We have seen in more recent history how our ancestors did 'go backwards' in the Dark Ages. When the Roman Empire withdrew from Britain, their elaborate brick-built, centrally heated houses were abandoned for the simple mud and stick houses of Medieval Britain and the warm air central heating facility was lost for nearly two thousand years.

So if we had houses in the past why did we go back to mud huts

Are we seeing the same thing in the Orkneys?

The only evidence archaeologists have of habitation is post holes in the ground. Most post holes would make fine square buildings, but this has always been dismissed since roundhouses were first defined some 60 years ago. The problem for archaeology is that once a discovery is defined in detail (usually via a paper or book on the subject, interpreting finds at a particular site) archaeologists apply it to all such features in their future fieldwork.

17000 BCE	10000 BCE	You are HERE
Ice Age Ends	Mesolithic Age Begins	

Are we seeing the same thing in the Orkneys?

Therefore, after Gerhard Bersu identified a roundhouse at Little Woodbury in the 1930s, this type of construction was expected to be the standard in all prehistoric sites throughout Britain. But there is a problem with this assumption. The houses at Little Woodbury were probably built in about 600 BCE. We know from the writings of the Romans that when they invaded a hundred years later, the society they conquered was hierarchical. That means that there were kings and servants, who one would imagine lived in different types of houses.

Now I understand that it is hard to believe that archaeologists are this 'prescriptive' in their attitude. But this is a real example of the problems that, if not challenged end up as perceived archaeological truth.

The Neolithic settlement at Brzesc Kujawski was discovered in 1933 by farmers digging gravel from deposits beneath their fields on a low ridge of land bordering Lake Smetowo. While digging, they found artefacts and skeletons. Luckily, an archaeologist named Konrad Jazdzewski (1908-1985) was working nearby, and when he learned of these discoveries he came to investigate. He immediately recognized that this was potentially an important find and began excavations. Over the next six years, he cleared the topsoil from more than 10,000 square meters, exposing one of the largest Neolithic settlements discovered before World War II.

Jazdzewski noticed that one of the most apparent Lengyel features at Brzesc Kujawski was the long narrow trenches dug into the clay and gravel subsoil, sometimes reaching a meter or more below the surface. These trenches formed trapezoidal outlines 20 to 30 meters long, 5 to 6 meters wide at one end and 2 to 3 meters at the other. Clearly, these were structures of some sort because there were indications that the trenches had held upright posts. Among these trapezoidal enclosures were large pits with very irregular bottoms dug into the clay subsoil".

So far so good then - now read the next extract - you couldn't make this up if you wanted too!!

"At the time, the prevailing belief was that Neolithic people lived in the pits, which were thought to have been roofed over with flimsy shelters. But what were the trapezoidal post structures? Archaeologists who had recently excavated Linear Pottery post structures at Koln-Lindenthal in Germany had proposed that they might have been barns or granaries. They could not imagine people living in them."

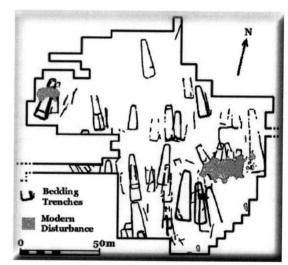

The Brzesc Kujawski Site

timeline

4500 BCE	2500 BCE	800 BCE	0-400 AD	1-2000 AD
Neolithic Age Begins	Bronze Age Begins	Iron Age Begins	Roman Period	Written History

Archaeologists (because of their inadequate and closed minded training, could not believe that these primitive hunter-gathers could build a house, so they told everyone that they lived in the ditch with an animal skin as a roof (no doubt attached to the perfectly secure wooden post).

It beggars belief, but it's absolutely true!

Even today they are suggestion that they are barns or granaries and people could not live in them - Is this because archaeologists have portrayed these people as 'fur covered primitives' that could only live in round mud huts, like our African ancestors - but it is a step up from living in a watery pit I guess!

"But one of Jazdzewski's workers remarked that if he had to live in one of the muddy clay pits, he would break his legs slipping around in it. Jazdzewski concluded that the Lengyel timber structures at Brzesc Kujawski really were Neolithic houses and that the pits served some other purpose."

IT'S CALLED A MOAT YOU IDIOT!!

A 'worker', some manual labourer had to break the news to Jazdzewski (Konrad Jażdżewski (1908–1985) a Polish professor of archaeology at the University of Łódź. that his ideas were plain 'NUTS' and the result of closed minded stereotypical twaddle. Its amazing to me, that 70 years later nothing has changed in the archaeology field and to prove it look at the new modern interpretation of the site.

"Eventually this view prevailed, and archaeologists now know that the big pits in fact were the places where clay was dug for plastering the walls of houses built with timber posts set into foundation trenches. At Brzesc Kujawski, more than fifty such houses have been found, both during Jazdzewski's excavations in the 1930s and during further excavations by Ryszard Grygiel and Peter in the 1970s and 1980s. They are oriented along an axis running northwest-southeast, with the wide end toward the southeast. The reason for this orientation of the houses or for their trapezoidal shape is not clear. Many of their outlines overlap, indicating that they were built and rebuilt at different times. Burned clay plaster in the filling of the foundation trenches indicates that a number of the houses were destroyed by fire. The nearby clay pits were filled up with debris, animal bones, charred seeds, and artefacts like broken pieces of pottery. Other pits were used for storage or as the locations of workshops" [15]

Clay pits for god sake!! Let's see clay foundations and rain, what do you get??..... it's the blind leading the blind!!

So Gerhard is probably right – the mud hut roundhouses so favoured by reconstruction archaeologists are typical dwellings – for some in that society, but not all; particularly not for tribal chiefs, princes and holy men and only from the Bronze age onwards 2500 BCE - 64 AD, which still leaves many possible buildings missing.

It's called a moat you IDIOT!

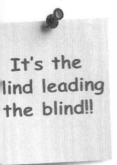

Post Holes

When you look at any prehistoric site, you find post holes; not just some, but sometimes hundreds. Archaeologists try to 'join the dots' and outline something familiar, but the chances of getting it right from just an outline of dots are realistically zero. Fortunately, later prehistoric roundhouses are much more identifiable, but the constructions of the Neolithic and Mesolithic are almost impossible to identify using today's techniques (or lack of them!).

The reason for the confusion is that most archaeologists seem not to understand why you dig a post hole! Structural archaeologist Geoff Carter has spent many a long year trying to educate archaeologists on the merits of post holes and why they are dug. On his web site, he concludes that 'thousands of disregarded postholes' are 'tucked away in reports as unphased.' He continues, 'it is little wonder archaeology made up a simpler story [of mud roundhouses] that was easier to understand'.

Quite simply, if you just want a stable post which does not wobble when pushed by wind or people, you use a stake, usually with a large spike on the end to cut through the soil – as you would do when erecting or mending a fence. The stake is simply driven into the ground, making a hole that narrows to a point. But you would only dig a hole and create a stable base, if you were going to place a lot of 'weight' directly on top of the structure.

A posthole is a hole dug to accommodate a timber post, which would typically be used to support a load acting vertically upon it, usually as part of a larger structure with multiple post foundations. For this reason, the top of the post would be narrowed to form a 'tenon' so it can be jointed into a corresponding hole, or mortise, in a horizontal timber. The horizontal timber will link it to other posts, and the weight of the structure will be spread evenly between several posts. Posts are least suited to loads being applied from the side, so in this situation, driven stakes, which fit tightly into the ground, would be used. This produces a feature archaeologists refer to as a 'stake hole', the ancient equivalent of a hole left by a modern sharpened fence post driven into the ground. Fences are not very heavy, but they do get pushed sideways by animals and even by the wind, making stakes more

Driven

Post

Stake

Dug

Stakehole

Posthole

4500 BCE	2500 BCE	800 BCE	0-400 AD	1-2000 AD
Neolithic Age Begins	Bronze Age Begins	Iron Age Begins	Roman Period	Written History

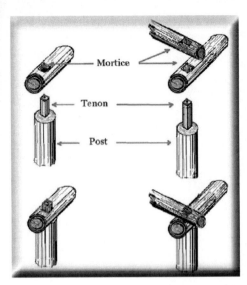

appropriate in these circumstances.

Stakes are designed to sink into the ground under pressure, whereas posts are not so designed; the terms, and the usage, are not interchangeable. Usually, the builder does not want a post to sink, since it is supporting the structure. However, pointed timbers driven a very long way into soft ground like marshes or lake bottoms can be used as foundations; in this context they are known as piles.' We have already seen mortise and tenon joints in the Archaeology section of this book, as Stonehenge's Sarsen stones were dressed with mortises and tenons before they were erected. This evidence proves that, prior to the building of Stonehenge, the mortise and tenon joint was very familiar to prehistoric man, and he must have used it in his domestic wooden structures.

If you have the ability to use this type of joint, and you understand the mathematics of weight distribution (hence the use of posts rather than stakes), then you can do two things. You can place very large roofs on your buildings; we can calculate the size of the roof space against the number of timbers required to support it. You can also build structures with multiple 'floors'. Man has always been fascinated by height, either climbing it or building it; our other ancient monuments had great foundations, not for a single storey enclosure, but to build to the sky.

Is it beyond imagination that at a time when our ancestors were building the largest man-made construction in Europe (Silbury Hill), they might also have built multi- storey monuments into the sky? In fact, I've built one myself in my garden for my children; it's called a tree house, didn't take long and it's over 10 feet high and I built it without any assistance. But don't ask me to build Stonehenge, for I wouldn't know where to start. So, if can build a construction above the ground on my own, why have archaeologists never considered that they were capable of multi-story buildings, even when they know from the joints used at Stonehenge they had the knowledge and technology?

These post holes can be reinterpreted to produce long houses as seen in the medieval period, which used the

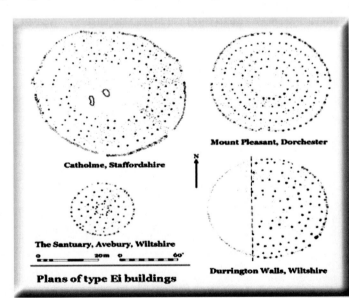

Catholme, Staffordshire

Mount Pleasant, Dorchester

The Santuary, Avebury, Wiltshire

Durrington Walls, Wiltshire

Plans of type Ei buildings

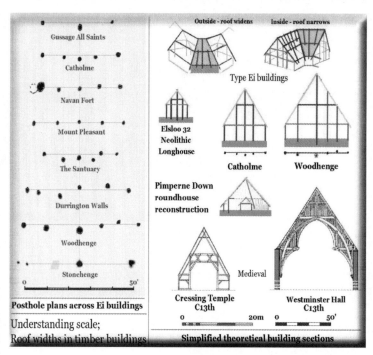

Posthole plans across Ei buildings

Understanding scale;
Roof widths in timber buildings

Outside - roof widens   Inside - roof narrows

Type Ei buildings

Elsloo 32 Neolithic Longhouse

Catholme   Woodhenge

Pimperne Down roundhouse reconstruction

Medieval

Cressing Temple C13th

Westminster Hall C13th

Simplified theoretical building sections

same woodworking techniques. Such floors could even be part of the roundhouse structures we have already found, that are currently identified as 'mud huts'. This, clearly, is an advanced civilisation that could build not only monuments of stone but also multi-floored houses or towers, such as Woodhenge.

Boat Houses and Crannogs

Remember, this was a nautical civilisation that travelled every day by boat, due to the thick forest and dangerous animals, and had marked trading routes (following long barrows).

That being the case, why would they wish to live on the land?

There are still indications of this aquatic dwelling style in Britain; they are called crannogs. These are houses that are built on the groundwater, connected to the land (if necessary) by bridges. The most important aspect of these houses is their accessibility to the boats the families would use. These type of homes existed up to the 17th century in Scotland, and if you look at Durrington Walls you can see that not only was it a natural bay for houseboats, but there are also outlines of five crannogs on the shoreline.

Recently, in Star Carr, archaeologists have claimed to have found the oldest house in Britain – its date, amazingly enough, is 8500 BCE. This is the same date we have suggested for the first phase of Stonehenge, including the three mooring posts in what is now the car park, but was then the shoreline of the Stonehenge peninsula. What the archaeologists found at Star Carr was a crannog, but more important was the discovery of the 'planking' that formed the walkway to the roundhouse, for they discovered the earliest known evidence of plank splitting. This is a process in which logs are split with wedges along the grain, to obtain flat thin wooden planks like those we use in construction today.

This find is significant, as it begs the question: why bother?

If you are making a walkway, it can be just as effective to use whole logs rather than split wood. The plank-splitting process would take more time and greater expertise to produce the desired effect. Well, logs are solid, and are hardier than split wood because they are thicker, so they would last longer. However, wooden slats are flatter and easier to walk on. If you're carrying goods and cargo, it's practical to use planks for a walkway over groundwater – they might save you from falling in!

What we see here is a high degree of expertise and technology that was thought to be more Iron Age than Mesolithic (a difference of about 7,000 years, incredibly enough), but the greater revelation is that planks could have been used for other purposes. If we are looking at a nautical civilisation that possessed not only wood-splitting skills, but also mortise and tenon joints as seen at Stonehenge, then it's not a quantum leap to suggest that they joined planks together to make large boats, rather than the dugout or leather canoes archaeologists had previously believed to be the limit of Mesolithic ability.

Crannog

If you were a master craftsman and you had ample free wood to use, where would you want to live: on land in a mud hut with the animals, or on the water on a boat of your own – away from dangerous predators?

History in other regions of the world have shown that mankind does not automatically jump from nothing to wooden boats, they use other materials, such as water reeds, which grow in abundance anywhere were shallow water lays, to form by just gathering and tying small single man boats. These reed boats obviously do not survive the passage of time, but yet we have found boats of reeds dating back to 5000 BCE in Kuwait and from inscriptions on temple walls we know they were common in ancient Egypt. Strangely, archaeologists are aware that our ancestors had access to reeds, but only used them for the roofs of the 'mud huts' they occupied.

If you can gather reeds to waterproof the roof of your round house - you can also make a boat!

The discovery of this water-based lifestyle changes our perception of prehistoric man, from a dirty hunter-gatherer dressed in fur to a boat-dweller living in a stable, idealist society with ample food and water. We should never forget the everyday essentials of life - not just food and water, but sanitary requirements too. Water is not only a great way of

Star Carr - planking 8500 BCE

17000 BCE   **10000** BCE

Ice Age Ends   Mesolithic Age Begins

Ancient Philippine Balangay

washing your clothes, but also of removing waste products and rubbish. Why dig a hole and bury waste in a pit for the insects or wild animals, when you can just dump it overboard?

One of the greatest archaeological mysteries has always been the lack of evidence for Mesolithic civilisation. We have always wondered why there is so little evidence of human habitation before the Iron Age, and clearly now we see houses and occupation aside from the occasional cave such as Cheddar Gorge. In some locations, flakes from stone knapping have been found, and it was imagined that these sites were ritual feasting locations for the nomadic Mesolithic tribes. Undamaged tools and trinkets discovered at the end of walkways were interpreted as sacred offerings to the water gods.

This is, of course, complete bunkum!

These items simply fell off the boats that were the Mesolithic people's homes, and could not be retrieved. By the same logic, you don't really pay homage to the great god of the sofa by offering him your small change! If you spend a lot of time on a boat, you get used to looking for a harbour when night begins to fall. The simple reason is that you need to meet some basic requirements in life. Firstly, you need to stock up with provisions. You may be sailing on the water, but you need to stop and obtain fresh water for drinking and cooking. You may also need some other foods as you will sometimes require or desire more than fish for your diet.

The second requirement is shelter. Boats don't do well in storms, and a harbour is the safest way of riding out any bad weather. The third requirement is company. Again, this is a basic human habit, and people who sail boats (particularly around the Mediterranean) always end up in a harbour at night with other people, as we are a species of social animal that loves to communicate and trade.

Harbours

We have found 3 harbours in this region (as seen in our earlier case studies). The one with the clearest evidence is Durrington Walls; as we have seen in our case study, it does not have a flat ground level. The semi-circle has a distinct 'V' shape, making it impractical to build houses there – but it is a perfect harbour. The banks were built up around it to protect the boats from the wind and rain, then crannog-like structures were built at the groundwater's edge to moor the boats and provide accommodation.

Next to the mooring site at Durrington Walls is the famous Woodhenge. Wooden henges were quite common in prehistoric times, and their purpose has always been marked down as religious or ceremonial –

timeline

| 4500 BCE | You are HERE | 2500 BCE | 800 BCE | 0-400 AD | 1-2000 AD |
| Neolithic Age Begins | | Bronze Age Begins | Iron Age Begins | Roman Period | Written History |

'don't know', in layman's terms! Lots of pits with animal bones have been found here, and burnt hearths without any associated houses – so, to archaeologists, they are ceremonial feasting stations! I suspect that these were just markets or trading stations where people congregated at night to eat, drink, talk and trade. Woodhenge is thought to have been roofed; what I believe is that it was more likely a tower with a roof, as it was next to a harbour and we know from ancient written books that this was common place.

Maybe Woodhenge even had a tower with a fire at the top, to signify that the site was open for business, which would have shone like a beacon at night to attract all those who needed food and shelter. Today we think of lighthouses as serving to warn boats of possible dangers, but they could have originated in the Mesolithic to attract boats to harbours. The earliest recorded lighthouse was built in about 300 BCE, on the island of Pharos at Alexandria in Egypt. I see no reason why such a monument couldn't have existed before that time. It is a fact that most lighthouses are situated on islands or at the end of peninsulas.

To prevent the wooden tower burning down, all you would need is a flat roof with a stone-lined fire pit under the hearth. The oldest surviving lighthouse in Britain is a Roman ruin at Dover. In antiquity, a lighthouse functioned more as an entrance marker to ports than as a warning signal, unlike lighthouses today.

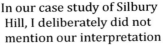

Alexandria Lighthouse - the first one?

In our case study of Silbury Hill, I deliberately did not mention our interpretation of the true purpose of this largest man-made object in Europe. I thought I'd save that fact to discuss here, because we believe that the hill was one the world's first ever lighthouses (Woodhenge was the other).

Silbury Hill Light House

The height, size and shape of the hill allows the occupants to place a massive beacon of fire on its flat top, which would have been seen for hundreds of miles across Mesolithic Britain. Avebury, as we have seen in the case study, was another massive inland port with islands and a gigantic stone circle where people met and traded; it is, again, a natural harbour, as it's at the end of a waterway. When you consider the lack of archaeological evidence under the mound, and the groundwater that surrounded this eccentric flat-topped construction like an island, it's no wonder that the

17000 BCE   **10000** BCE

Ice Age Ends   Mesolithic Age Begins

purpose of Silbury Hill has baffled archaeologists. There were no dead bodies, no gold, no treasure of a great king – it's much more valuable than that, for it was a beacon that shone to gather together a great civilisation in what became the dawn of mankind.

Wood

Before we move on to discuss trade, we need to look at the ease of use and access to wood. Archaeology has given us strange ideas about the past and the people who lived in prehistoric times. We have disproved the idea of hundreds of men pulling large stones over a grassy Salisbury Plain, and replaced it with a vision of a nautical society floating stones down rivers by attaching them to large wooden rafts. Wood would be plentiful; it really does grow on trees! But again, we are left with the strange idea that prehistoric man spent hours, days and weeks cutting down these trees, which is not the case.

If you want to fell a number of trees in a short period, you don't chop them down one by one, because it takes too long and is too exhausting. The easiest way of felling a tree is to burn its base with fire. Even the mightiest tree, such as the metre-wide trees found at the Stonehenge mooring, could be burnt down over 36 hours of intensive burning. This process allows you to fell several trees at the same time; all you need to do is maintain the fire, and make sure you aren't under the trees when they become weakened and begin to fall! The downside to this method is that occasionally, in the dry season, the fire may spread out of control and burn through massive swathes of forest. This may give us a further clue as to why such vast areas of forest were cleared in the Mesolithic and Neolithic periods; perhaps they were burned not for farming as was first believed, but rather in the earliest industrial environmental accidents.

We know that this burning process took place, as charcoal has been found at the bases of some postholes where the builder failed to cut off all the burnt material before erecting the posts. Fortunately, because charcoal does not rot, we can still carbon date the postholes (as at the Stonehenge car park).

Trade

Man has traded almost since the dawn of his creation. The first evidence of trading has been dated to 150,000 BCE – long before the Mesolithic Period, so we can be quite assured that our ancient civilisation traded. When prehistoric man moored his boat in that safe haven, guided by the lighthouse to show it was open, the chances are that others would be doing the same. If he had a surplus of fish, tools or trinkets he would probably end up trading.

There is no indication that money existed in those times, so we can assume that the barter system was used to trade goods – "I'll swap you 10 fishes for that watery stuff with an added kick", or "I'll build you a boat in exchange for that wonderfully polished flint tool". Bartering is a unique system that is alien to us as we live in a monetary system; history tells us that societies that used bartering systems had markets, and rules governing trade. This indicates a civilisation of great sophistication for, as we have now proved, the prehistoric Britons had what is known as an organised marketplace.

Floating Market - still found in Asia

Nonmonetary societies operated largely along the principles of gift economics. When barter did in fact occur, it was usually between either complete strangers or would-be enemies. While one-to-one bartering is practised between individuals and businesses on an informal basis, organized barter exchanges have developed to conduct third party bartering. The barter exchange operates as a broker and bank and each participating member has an account which is debited when purchases are made, and credited when sales are made. With the removal of one-to-one bartering, concerns over unequal exchanges are reduced[16].

Are we seeing, in these harbours and lighthouses, a civilization with such a sophisticated bartering system that they had created the world's first trade. A trade or barter exchange is a commercial organization that provides a trading platform and bookkeeping system for its members or clients. The member companies buy and sell products and services to each other using an internal currency known as barter or trade dollars. Modern barter and trade has evolved considerably to become an effective method of increasing sales, conserving cash, moving inventory, and making use of excess production capacity for businesses around the world. Businesses in barter earn trade credits deposited into their account instead of cash. They then have the ability to purchase goods and services from other members, using their trade credits; they are not obligated to purchase from the person they sold to, and vice versa. The exchange plays an important role because it provides the record-keeping, brokering expertise and monthly statements to each member. Commercial exchanges make money by charging a commission on each transaction, either all on the buyer's side, all on the seller's side, or a combination of both. Transaction fees typically range between 8% and 15%.

The 'middle man' of the exchange received a percentage of the value of any goods traded. This would have

17000 *BCE*   **10000** *BCE*

Ice Age Ends   Mesolithic Age Begins

allowed him to barter for materials and a workforce to build the harbour and lighthouse. Or did his society understand that such a facility was necessary, and volunteer to build and use it together for the common good (making them prehistoric socialists!)? Either way, we have in one short chapter taken the archaeologists' half-naked hairy Mesolithic man living in a cave (remember, mud huts came some 5,000 years later) and shown that in fact he was a boat-sailing entrepreneur with a marketplace to sell his goods and services. Someone is sadly deluded!

If we are right about this lost advanced civilisation, surely there must be other evidence of trade?

Note in margin: Are these nonuments to result of the niddlemen?

Tools

It's almost impossible to date a flint tool, unless it's found 'in situ' and we can date the layers above and below. It seems reasonable to assume that the most primitive tools are older than the most sophisticated, but it doesn't appear to work out that way when you look at what evidence has been discovered. The first thing you will notice about Mesolithic tools is that they are sharper than the later Bronze Age tools! The counter to that is that they are more brittle. Were these tools made for the mass market? Are they are the first disposable tools in history?

Why would you make such a tool?

Because, as any entrepreneur will tell you, that's how you stay in business – by keeping people coming back for more. So, Mesolithic tools were more efficient, but Neolithic tools started to feature polished blades; these not only looked good, but could be re-polished and sharpened. The next strange item in the prehistoric tool kit is the stone axe – this was an essential item for survival in prehistory. There are three types of axe which have been dated to this period. First came the rough and ready axe with flint flakes bound to a wooden handle, very sharp, very practical.

Mesolithic microlith tools

Then came an axe which was highly polished and attached to the same type of handle; it is less sharp, but more robust, and looks fantastic. These axes show the hours of work needed to make them smooth. Archaeologists call these ceremonial axes as they show little signs of wear; alternatively, it's possible that the axes were simply very well cared for and had all the dents removed on a regular basis – a little bit like washing and polishing

timeline

4500 BCE	You are HERE	2500 BCE	800 BCE	0-400 AD	1-2000 AD
Neolithic Age Begins		Bronze Age Begins	Iron Age Begins	Roman Period	Written History

your car, it's not a ceremonial thing but a matter of pride. The third axe type is a complete mystery, and gives us an indication of the sophistication of the period. These axes were not only highly polished, but had a 'mace' handle cut into the stone. This may not seem such an incredible achievement at first sight, but if you've ever tried to drill a hole in a piece of granite, you will know just how hard it is. A civilisation that could achieve this using only basic hand tools deserves our respect.

I tried to use an electric high-speed hammer drill with a diamond head to cut a hole just 10mm into such a piece of granite. After applying the full force of my 14 stone body in the attempt, which resulted in frustration and only a tiny dent in the surface after several minutes, I gave up and called in an expert with an even more powerful machine. The truth is, no one knows, and archaeologists don't want to ask too many inconvenient questions as according to them these primitives spent all day hunting food and living in mud huts – get the picture?

To drill these types of holes, you must have either metal or a tool called a 'bow drill' (or, for the larger holes in granite, a 'pump drill'). The problem with these instruments is that according to current theories they were not used until 3000 BCE, in Egypt. So, either way, this prehistoric British civilisation had the skills and technology with which we credit the ancient Greeks and Egyptians some 5,000 years later. Is it possible that our ancestors used a simpler method?

What we must keep in mind is that equipment such as drills would allow the level of sophistication in other areas, such as furniture, to evolve above all expectation. The standard of living would progress from mud huts to chairs and tables, and the evolution does not stop at housing; drills can also be used for other purposes, such as medicine.

Medicine

Dentistry has been found in a dozen or so human teeth dated to the Neolithic Period. It appears a bow drill was used to cut out decay and relieve pain as the drilling is extremely accurate. Moreover, this is an unbelievably sophisticated method; a tooth extraction is much simpler, does not require an expert, and is just as effective. That drilling was carried out instead of extraction indicates the strength of medical care in prehistoric society. Further medical evidence can be seen in one of the great untold mysteries of prehistory. Highly developed surgery methods of the Neolithic Period have been discovered over recent years.

Scientists excavating a Neolithic tomb at Buthiers-Boulancourt, near Paris, found that its occupant had undergone a surgical amputation. The elderly man buried in the tomb had his left forearm carefully removed about

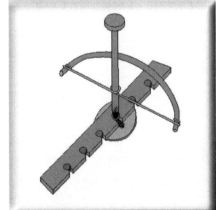

By using a Bow Drill

'Expert' Neolithic Amputation

6,900 years ago, demonstrating that our ancestors had quite a remarkable degree of medical knowledge. The French National Institute for Preventive Archaeological Research reports that the patient is believed to have been anaesthetised with pain-relieving plants; the conditions were aseptic enough to avoid infection; the cut was clean, and the wound was treated with herbal medicine (it's amazing what you find when you fund archaeology correctly!).

Intriguingly, the cut was made just above the 'trochlea indent' at the end of the bone, indicating that the surgery was carried out by someone with a high degree of medical expertise. This was just one example; two other Neolithic amputations have been discovered in Germany and the Czech Republic. So, is it possible that this man travelled to Stonehenge for his treatment, or had advanced medical knowledge spread from Britain to Europe by 5000 BCE, some 3,000 years after Phase I of Stonehenge's construction?

We know that the man was well travelled, as one of his stone tools came from the stone range of the Ardennes in Belgium (at that time, Belgium would still have been connected to 'Doggerland', as the English Channel had not yet fully formed).

Neolithic Amputation of the arm - undertaken with 'expert' skill But it doesn't end there; here is more evidence of a society that was 'before its time' according our present view of history. Trepanation is a surgical operation that involves the removal of a rectangle or disk of bone from the skull. The section of bone may be extracted with flint or metal blades by drilling a series of small holes, making intersection incisions, or scraping through the bone.

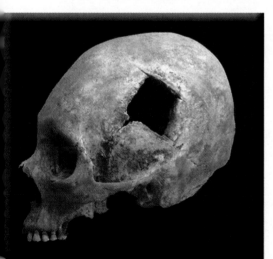

The Buthiers-Boulancourt burial contained the well-preserved skeleton of a man who died at roughly 50 years of age. The radiocarbon date for his bones was 5100-4900 BCE, about 2,000 years before current theories say that Stonehenge (the centre of the most technologically advanced society in European prehistory) is supposed to have been built. Two trepanations had been carried out on this man: one to the front, measuring 2.5 inches by 2.6 inches, and one to the top, measuring 3.7 inches by 3.6 inches. The frontal operation scar was completely healed, and the larger surgery wound on the top of the skull had partially healed, indicating that he survived both operations successfully.

timeline

4500 BCE	You are HERE	2500 BCE	800 BCE	0-400 AD	1-2000 AD
Neolithic Age Begins		Bronze Age Begins	Iron Age Begins	Roman Period	Written History

Another interesting aspect of this story that was missed by most commentators was the age of the man found. He was 50!

According to experts, the life expectancy of a Neolithic man was 20-25. The Romans were not expected to live more than 30 years. In fact, a life expectancy of 50 was not reached until the 20th century, so what kind of society could have such long life expectations? And this is not an exception; judging from the bones found in Britain so far, the average age of a Mesolithic man at death was 35. Is it possible that a civilisation had developed that was so sophisticated that they built massive stone monuments, and were able to cure the sick with health spas and surgery? If so, did our French man travel to Stonehenge for his operation? If he did, is there any evidence that other people travelled these vast distances to be cured of their ills?

The Amesbury Archer is one of the most famous stories about Stonehenge that has come out in recent years. This man, who was suffering from a bone infection, travelled from the Italian Alps to Stonehenge, but died and was buried with a selection of grave goods that amazed archaeologists, as they gave an insight into the life of a man who lived some 4,500 years ago.

So the Archaeologists said he must have walked. Unfortunately, he was missing his left knee cap, so he must have hopped more than 800 miles to get here!

The Amesbury Archer must have travelled by boat to Stonehenge. This means that this trading route had already been well established, as most people would have to take time to consider and prepare for such a journey. The grave goods found on the Archer show that he was a craftsman, so he would have traded his craft in exchange for food, water and accommodation during his trip to Britain. He must have also taken his family, as he was buried with all his worldly goods, which would have been very unlikely if a stranger buried him.

Moreover, the Amesbury Archer was about 35 to 45 years old, 10 years older than the life expectancy of an Iron Age man, clearly showing that the treatment found at Stonehenge was successful even for his crippling illness.

Farming

Traditionally, the Neolithic Period is famous for the birth of civilisation and farming. We need to spend a couple of moments considering this aspect of life. The classic classroom bunkum would normally go like this:

The change from a hunter-gatherer to a farming way of life is what defines the start of the Neolithic or New Stone Age. In Britain the preceding period of the last, post-glacial hunter-gatherer societies is known as the Mesolithic, or Middle Stone Age. It used to be believed that the introduction of farming into Britain was the result of a huge migration or folk-movement from across the Channel. Today, studies of DNA suggest that the influx of new people was probably quite small - somewhere around 20% of the total population were newcomers.

So the majority of early farmers were probably Mesolithic people who adopted the new way of life and took it with them to other parts of Britain. This was not a rapid change - farming took about 2,000 years to spread across all parts of the British Isles. Traditionally the arrival of farming is seen as a major and rapid change sometimes called the 'Neolithic revolution'. Today, largely thanks to radiocarbon dates, we can appreciate that the transition from hunter-gatherer to farmer was relatively gradual. We know, for example, that hunters in the Mesolithic 'managed' or tended their quarry. They would make clearings in woodland around sources of drinking water, and probably made efforts to see that the herds of deer and other animals they hunted were not over-exploited.

The switch from managed hunting to pastoral farming was not a big change. The first farmers brought the ancestors of cattle, sheep and goats with them from the continent. Domestic pigs were bred from wild boar, which lived in the woods of Britain. Neolithic farmers also kept domesticated dogs, which were bred from wolves. It is probable that the earliest domesticated livestock were allowed to wander, maybe tended by a few herders. Sheep, goats and cattle are fond of leaves and bark, and pigs snuffle around roots. These domestic animals may have played a major role in clearing away the huge areas of dense forest that covered most of lowland Britain

Neolithic Farming House

This is quite a modern interpretation, taken from the BBC history web site. Radiocarbon dating is now leading to an adjustment of the date archaeologists had established for the earliest signs of 'civilisation' or farming, which was about the same time as Stonehenge in about 3000 BCE. This has now been slowly extended, as more evidence is gathered, to the late Mesolithic in 5000 BCE. Yet even with overwhelming evidence to the contrary, archaeologists drag their heels at the suggestion that the old dating system is simply wrong! In the quoted website article, the evolution of the domestic dog is placed with farming (as they believed it was associated with sheep farming) in the Neolithic. However, the earliest known domestic dog bones were found at another famous Mesolithic camp, Star Carr, dated to 7500 BCE.

This either means that sheep farming started 2,500 years earlier than archaeologists are telling us, or (as we believe) that dogs were domesticated long before farming arrived – which opposes the traditional view. At the end of the Ice Age, our ancestors walked the tundra tracking for food just like the Eskimo do today; they would likewise have used dogs to pull their sledges. These dogs were bred from wild wolves, which would have become domesticated in the course of time. This means that the dogs would have been passengers on the boats of the civilisation that built Stonehenge, as sheep had yet to be introduced. Farming was probably taken up in the Mesolithic Period, in about 5000 BCE. Initially, this would have started from ideas brought to Britain by traders from other, older civilisations within Europe.

Farming was not necessary in Britain, as our boat-dwelling ancestors had a plentiful supply of fresh fish and other seafood, but in a trading nation, supplies of exotic foods would have carried greater value than the abundant staple supplies relied on in the past. So tools could be purchased with some food supplies in this open marketplace. Later in the Neolithic Period, farming became a growing necessity as the groundwater tables throughout the country fell and life as a boat family became increasingly difficult. At this point, our ancient civilisation had a decision to make that would change the course of British history.

Our ancestors decided to leave and move South, taking their technology, culture and philosophy with them. Regarding this journey, I will go into further detail in a future book.

Conclusion

This is not the only time in our history that an established civilisation has come to the fore and then subsequently disappeared, to be replaced with a lesser form of society. The ancient Romans brought new technology to rural Britain in 45 AD with brick houses, central heating, glass windows, roads, wine, sanitation, art and literature (we'll forgive them for slavery for the time being!), Only to be replaced in less than 200 years with wood and mud huts, disease from poor sanitation, hunger from poor cultivation, and lots of beer drinking by horned warriors. We now refer to this as the Dark Ages within the Mediaeval Period that lasted about 1,000 years. As we are aware of such reversions in history, archaeologists should be more careful in their assumptions about how our world has developed. This book has revealed a new world of groundwater in the Mesolithic Period; it has also discovered that a civilisation of huge potential did in fact exist 5,000 years before conventional archaeology and history recognises their existence.

This knowledge fundamentally changes the history of Britain, and subsequently the world, because unless our prehistoric ancestors' technological and mathematical skills were completely lost, they must have passed down their skills to other generations and cultures. When we look at the history of the world, particularly if we seek to find out where the ancient Greeks and Egyptians first obtained their engineering and mathematical knowledge, we find that we have been left with an incomplete history of their origins. In some instances, this practical knowledge seemed to 'appear' suddenly, without any evidence of prototypes

17000 BCE   **10000** BCE   You are HERE

Ice Age Ends   Mesolithic Age Begins

or of failures in the course of its development.

But there are accounts of a great civilisation that may have influenced ancient Greece. For example, great writers such as Plato referred to such a nation with advanced technology and knowledge, which influenced their society through trade. Within this book, we have proved that a great nautical civilisation did exist that was capable of building monuments of such magnitude that they still lie on Britain's hillsides some 10,000 years later. This civilisation traded throughout Europe, and had knowledge of medical procedures and techniques not seen again for another 7,000 years in Britain. The reputation of their medical treatments enticed the sick and disabled to travel across the known world to bathe in the pure waters of a Mesolithic hospital.

In Plato's account, this great civilisation was a 'naval power' lying in front of the 'Pillars of Hercules' (the opening to the Atlantic Ocean from the Mediterranean). It conquered many parts of Western Europe and Africa, 9,000 years before the time of Solon (i.e. in 9500 BCE). Amazingly, this is approximately the same date as one of the four gigantic metre-wide mooring posts was being placed on the shoreline of a peninsula on what is now known as Salisbury Plain. This mooring post, which has been carbon dated to 8500 BCE ±500, was the beginning of a construction that was to become the greatest and oldest surviving monument of the prehistoric world: Stonehenge.

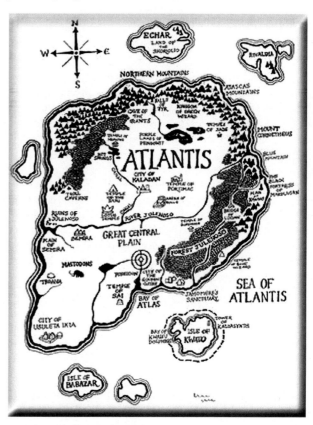

In our next book 'Dawn of the Lost Civilisation' we will trace the movements of this great society that started its journey in the Caspian Sea and finally travel to all four corners of the world, where there DNA lays testament to being the forefathers of our civilisation, for Plato was referring to the great 'lost civilisation' known to the ancient Greeks as:

Atlantis.

timeline

4500 BCE	2500 BCE	800 BCE	0-400 AD	1-2000 AD
Neolithic Age Begins	Bronze Age Begins	Iron Age Begins	Roman Period	Written History

Epilogue - Ancient monument to the Lost World of Atlantis

If the builders of Stonehenge were the lost civilisation of Atlantis, where did they originate from and why did they come to the Salisbury plain to construct their monument?

We have shown that after the last ice age the 'great melt' flooded the landscape of Britain for over ten thousand years. This flooding started a 'chain reaction' and as a consequence the sea level slowly increased and vast areas of land were lost including a land mass known as 'Doggerland'.

Throughout the 19th century, oyster dredgers working the shallow waters off the north east coast of England recorded frequent finds of animal bones caught up in their nets. These discoveries became a regular occurrence as the fishing technology increased and the trawlers at a later date and in deeper waters of the North Sea also found traces of civilisation and a lost continent. Sadly, the location were rarely recorded with any degree of accuracy; this material appeared to come from a number of areas within the North Sea.

One area where the greatest number of finds were was discovered is known as the 'Dogger bank' which lies just 90km – 110km (60-70 miles) from the coast of the British Isles. This shoal (a shoal, sandbar , sandbank or gravel bar is a somewhat linear landform within or extending into a body of water, typically composed of sand, silt or small pebbles.) rises about 45m (150 ft) above the North Sea bed. To the north it plunges into deeper water and forms a subterranean plateau covering 17,600 sq. km. (6,800 miles) with a maximum dimension being 260km (160 miles) from North to South and 95 km (60 miles) from East to West. Over time the number of finds reduced as the same area was dredged day after day and any artefacts sitting on the surface would have been scooped up and either returned as a curiosity (or in a majority) most of the time just thrown back in a different location.

Even so, items such as; bear, wolf, hyena, bison, woolly rhino, mammoth, beaver walrus, elk, deer and most importantly horse have been collected. This precious collection of findings gives us a fantastic insight to what Doggerland looked like, the environment that supported these animals and the climate of this unique area of the world.

It would be impossible to talk about Doggerland and its environment without understanding the clear connection to Plato's written references to the lost world of Atlantis. Over the years Atlantis has

Doggerland

17000 BCE   10000 BCE   You are HERE

Ice Age Ends   Mesolithic Age Begins

Plato

grown to be both a legend and the source of much science fiction. This is neither correct nor helpful in tracing the history of mankind, as it moves the debate from scientific observation to fantasy and the degrading of the most important time of our history.

Plato is a source of credible information for he is not a 'story teller' like some other historic writers, he is fundamentally a philosopher whose writings are still studied even now, some 2,000 years after his death at the most famous and prestigious universities throughout the world. This man is not prone to fantasy or exaggeration, his writings therefore must be accepted as true evidence that once in the distant past an ancient great civilisation did in fact exist and that they changed the course of mankind in ways which I believe we do not fully understand to date.

The next book in our trilogy traces the 'megalithic builders' from when they came out of Africa to where they landed in their boats on Doggerland following the herds of Animals as the ice caps receded and the food supply for these ancient ancestors moved north travelling to a new continent. They constructed a civilisation that used new stone tools and incorporated megalithic stone constructions into their society.

Consequently, we need to look at the probability that Plato's 'Atlantis' is a genuine reference to this land, as it is the oldest written source and may give us clues of how this civilisation lived and traded. Fortunately for us Plato gave some detail about this civilisation, such as how they lived and what they believed, which will allow us to compare what we know from landscape and archaeological finds and look for other areas of investigation the texts might reveal.

Plato's most famous line from 'Timaeus', a dialogue between Critias and Socrates, where 'Critias' tells a story he learned through his family about the Greek statesman 'Solon' whilst he was studying with the most scholarly of Egyptian priests during a visit to Sais in Egypt in about 590 BCE. The priests claimed to have access to secret records about a lost civilisation called 'Atlantis', which only they were allowed to read, for it was written on the pillars within their most sacred temple. Now Sais was one of the oldest cities in the old kingdom and the city's patron goddess was 'Neith', whose cult is attested as early as the 1st Dynasty, ca. 3100- 3050 BCE.

The Greeks, such as Herodotus, Plato and Diodorus Siculus, identified her with Athena and hence postulated a primordial link to Athens. Diodorus recounts that Athena built Sais 'before' the 'deluge' that supposedly destroyed Athens and Atlantis. While all Greek cities were destroyed during that cataclysm, the Egyptian cities including Sais survived. As we can see from this connection, the deluge has incredible importance to ancient civilisations, clearly indicating that any prehistoric civilisation that wanted to 'stay alive' would

timeline

4500 BCE	2500 BCE	800 BCE	0-400 AD	1-2000 AD
Neolithic Age Begins	Bronze Age Begins	Iron Age Begins	Roman Period	Written History

possibly build boats, not for some, but for everyone. Sadly, the city of Sais has been recently destroyed by farmers who used the house and temple mud bricks as free fertiliser for the fields – to this date the temple and its writings have never been found.

The most famous line from Plato's dialogue is "in front of the mouth which you Greeks say 'the pillars of Hercules' there lay an island which is much larger than Libya and Asia together" translated by W.R.M. Lamb 1925 or "in front of the straits which are by you called the pillars of Hercules; the island was bigger than Libya and Asia together" B. Jowett 1871

This single sentence has caused no end of debate about the location of Atlantis. Some suggest that 'the pillars' can refer to water flows, thus allowing the speculation (which is current) that Atlantis is a Greek Island. Others suggest (including myself) that the 'pillars of Hercules' is the mouth of the Mediterranean between Morocco and Spain. Now this is a case of translation and interpretation, the word 'mouth' is sometime called 'strait', in other quotations Plato refers to the Mediterranean Sea as "within the straits of Hercules".

According to some Roman sources, while on his way to the island of 'Erytheia' Hercules had to cross the mountain that was once Atlas (the Atlas Mountains are in Northern Africa overlooking the Mediterranean). Instead of climbing the great mountain, Hercules used his superhuman strength to smash through it. By doing so, he connected the Atlantic Ocean to the Mediterranean Sea and formed the Straits of Gibraltar. But the best evidence is in the name itself 'Atlantis' for Herodotus (an ancient Greek historian, 484 BCE – 425 BCE) in a time before Plato's writings calls the Sea outside the Pillars of Hercules the 'Atlantis Sea' (Cyrus, 557-530 BCE: Book 1). Moreover, even today we call it the Atlantic Ocean and in history c's and s's are commonly transposed.

So we are left with a clear understanding that Atlantis was in the Atlantic Ocean, but then come the next problem with this description "the island was bigger than Libya and Asia together" this is where most Atlantis claims fall flat. Libya was well known in Plato's time as a big country as it bordered the Mediterranean, but the reference to Asia cannot be the Asia we know as it was unknown to the old world and the Greeks, therefore the Asia that Plato was referring to is now called Asia Minor.

'Asia Minor (from Greek: Μικρὰ Ἀσία, Mikrá Asía, small Asia) is a geographical location at the westernmost protrusion of Asia, also called Anatolia, and corresponds to the western two thirds of the Asian part of Turkey. It is a peninsula bounded by the Black Sea to the north, Georgia to the north-east, the Armenian Highland to the east, Mesopotamia to the south-east, the Mediterranean Sea to the south, and the Aegean Sea to the west.'

The size of this 'island' is consequently a major problem for historians to date, as the only two island possibilities are the Caribbean in America or a continent that was once in the middle of the Atlantic Ocean that has disappeared without trace. Well the islands of the Caribbean are far too small and the trek across the middle of the Atlantic Ocean without landmass to guide the ships eleven thousand years ago would be too daunting to be truly feasible with 'Bronze Age technology' as Plato suggests.

This is why the search has failed to date and all various 'silly' hypotheses based on the Mediterranean make news headlines. If we look again at this passage and the exact wording of Atlantis we find something most

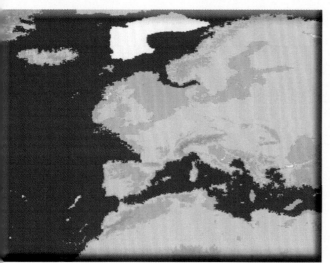

Plato's Atlantis - North European Peninsula

researchers have overlooked in the translation and it's the word 'island', the original Greek word is 'nesos' which can mean either island or peninsula.

If we are looking for a 'Peninsula' (which is a piece of land that is bordered by water on three sides but connected to mainland.) that is outside the Mediterranean, then there are only two possibilities - Africa or Europe. These are both outside the Pillars of Hercules and can be easily navigated by sticking to the shorelines. The African continent has shown no signs of any peninsula on its Atlantic side that has disappeared in the past 10,000 years - but Europe has!

If we look at a map of Europe at the end of the Ice Age, we notice that the water levels were about 160m lower than today, so much lower that extra coastlines are added to both Spain and France. But when we look at the British Isles we notice Britain has completely vanished. It has been replace a massive new land mass protruding into the Atlantic Ocean, for the English Channel, Irish Sea and the North Sea as we know them today has been replaced by a single land mass. Moreover, the land to the west of Ireland and North West from Scotland would have also been reclaimed from the sea.

This peninsula (which includes to the North East Norway, Sweden and Finland, to the east Denmark and the Baltic Sea) creates a continent about the same size of Libya and Asia Minor, which correlates to Plato's writings.

We know from our history that the rising of sea waters over the last 10,000 years has caused flooding that created the island nation we know today. But, do the writings contain any other information which will allow us to confirm this peninsula is the land mass Critias was talking about?

Plato adds "yonder (beyond the pillars of Hercules) is a real Ocean, and the land surrounding it may most rightly be called, in the fullest and truest sense, a continent" in this sentence the 'island' is turned into a 'continent' so this proves that the translation of 'nesos' is peninsula not island, and in today's terms, we are looking at a land mass that incorporates the British Isles, Scandinavia and the Northern European countries of France, Germany, Holland, Belgium, Poland, Netherlands, Denmark, Lithuania, Latvia, Estonia and the Baltic, North and Irish seas, that were at time land masses, which I call the North European Peninsula (NEP).

Therefore, do the other descriptions of Plato's Atlantis match this new continent?

timeline

4500 BCE	2500 BCE	800 BCE	0-400 AD	1-2000 AD
Neolithic Age Begins	Bronze Age Begins	Iron Age Begins	Roman Period	Written History

Where he does give us an indication of some other identification features for this lost continent is "and it is possible for the travellers of that time to cross from it to the other islands and from the islands to the whole of the continent". If NEP is correct, this could mean either one of two different continents. For about 2,000 years (between 10,000 BCE and 8,000 BCE) you could travel from the North West end of the NEP to the Faroe islands which were about five times larger than the islands we know today (due to the drop in sea levels) and then a short hop west to Iceland, which again, for the same reasons, was twice as wide as today, then finally over the short distance to Greenland and then America.

The coastal route would then allow you access to the 'New World' down the entire east side of America. The most interesting and controversial aspect, is that the Atlanteans would have discovered America 11,000 years before Columbus, which may seem farfetched but it will answer an evolutionary mystery that has confused geneticists and anthropologists for many years. This mystery involves the spread of the A- blood group in Northern American Indians and ancient skeletons that have European features and stone tool kits.

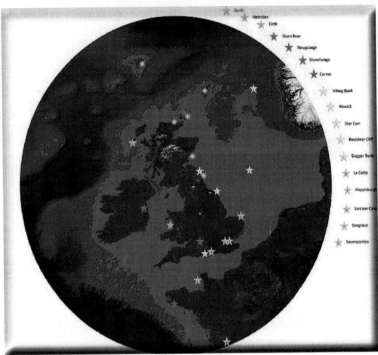

Atlantis (NEP)
- showing major Mesolithic sites

Other illustrations of Atlantis, in Plato's writings, can now be compared for further comparison such as the famous plain of Atlantis. For "The whole country was said by him to be very lofty and precipitous on the side of the sea, but the country immediately about and surrounding the city was a level plain, itself surrounded by mountains which descended towards the sea; it was smooth and even, and of an oblong shape, extending in one direction three thousand stadia, but across the centre inland it was two thousand stadia."

So a level oblong shaped plain that was surrounded by mountains which were by the sea and the plain which measured about 250 miles by 350 miles. The current North Sea (Doggerland) would comfortably incorporate a flat plain of that size and hence it sinking below the rising sea levels, for it measures 450

17000 BCE   You are HERE   10000 BCE

Ice Age Ends   Mesolithic Age Begins

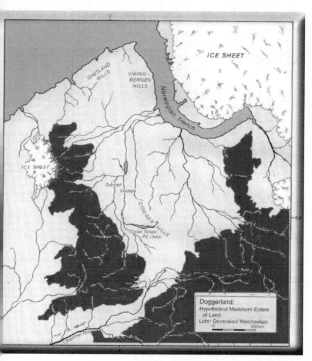

miles by 1200 miles. There are mountains in the North in both Scotland and Norway which 'descended towards the sea' not even taking into account the mountains of Doggerland in the north which is now the 'Viking' and 'Berger' sand banks between Shetland and Norway.

Another indication that we are talking about a peninsula or part of a continent is the line 'lofty and precipitous on the side of the sea' if this was an island would not ALL sides be on the side of the sea? Our newly discovered Atlantis has one side by the sea where the city and plain lays and that is in the north, the area we know now as Scotland, Shetland, Orkney across to Norway, where it always rains.

He continues; "This part of the peninsula looked towards the south, and was sheltered from the north. The surrounding mountains were celebrated for their number and size and beauty, far beyond any which still exist, having in them also many wealthy villages of country folk, and rivers, and lakes, and meadows supplying food enough for every animal, wild or tame, and much wood of various sorts, abundant for each and every kind of work." Modern sonar readings of the North Sea floor has shown this land of 'rivers and lakes and meadows' and with the mountains in the north the great plain would naturally face south.

Even when we search the description for the most obscure references to test this hypothesis we find correlation; "There was an abundance of wood for carpenter's work, and sufficient maintenance for tame and wild animals. Moreover, there were a great number of elephants in the island; for as there was provision for all other sorts of animals, both for those which live in lakes and marshes and rivers, and also for those which live in mountains and on plains, so there was for the animal which is the largest and most voracious of all." Plato is clearly referring to tropical animals such as Elephants and Lions and anyone living in Britain today would find this unlikely and therefore rule out this continent. But again, we found in the catches of North Sea trawler men the bones of Elephants, Mammoths, Lions and Tigers in the last 100 years in Doggerland making this scenario 'quite probable' and it would not take much of a stretch of the imagination to classify 'the largest and most voracious of all' as the ferocious Sabre-Toothed Tiger, which coincidently went extinct during the 'Atlantean period' 14,000 BCE – 4000 BCE.

timeline

4500 BCE	2500 BCE	800 BCE	0-400 AD	1-2000 AD
Neolithic Age Begins	Bronze Age Begins	Iron Age Begins	Roman Period	Written History

So we are looking at a civilisation of megalithic builders that once lost their 'homeland' and no doubt a great number of friends family and loved ones in the deluge. This being the case - how would we expect a civilisation to mark such a momentous occasion?

We are a civilisation that has had great losses through war and tragedy in the past, 9/11 in America we commemorated by memorial gardens and here in Britain the war losses are remembered by the cenotaph in London, where we come annually to remember the dead with a two minute silence.

How would future archaeologists view the cenotaph, if all written records are lost 10,000 years in the future - when the wind and rain have wiped the stone monument bear, but still resembling the stone pillar it once was?

We now know from previous chapters that Stonehenge was built about 8500 BCE as a place for the sick and dead, we also know that the site was remodelled some four to five thousand years later, roughly at the time that Atlantis finally 'disappeared'. What I am suggesting is that the new monument continued its commemoration to the dead, but not to individual dead but the death of Atlantis, which would explain how and why this historically massive construction was erected.

City of concentric circles - as found at Avebur

Furthermore, it will explain some of the features of Stonehenge we have not mentioned to date in this book. The popular view of Stonehenge is of a completely round monument with lintel stones completing the circle - but there is a massive problem, for not all the stones or the stone post holes are present. Some people have suggested that the monument is incomplete, but those individuals do not understand the monument and the reason for its construction.

A monument to the dead does not face the Summer Solstice Sunrise it faces the Winter Solstice Sunset or Sunrise to mark the shortest day - when light overcomes the darkness, the symbolism of life after death. That being the case, a monument to the dead should face either Winter Sunrise or Winter Sunset and Stonehenge faces the latter. Another reason to ignore the reference to the Sun is that the Sun is usually a representation of life and a completed circle, were as the Moon represents the dead and a crescent. So, the monument would be crescent shaped and hence the reason for the lack of two stones in the South West Quadrant.

So why the horseshoe crescent in the centre of the monument?

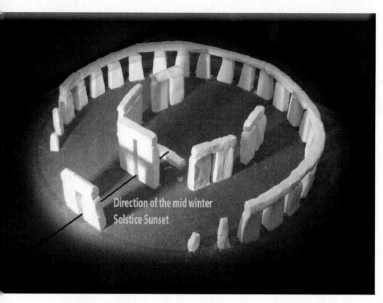

Direction of the mid winter Solstice Sunset

The crescent within the crescent faces the Summer Sunrise, this is very symbolic as it represents rebirth or reincarnation. It's a poignant message through the ages to us decedents of the Atlanteans, where they are telling us that their homeland maybe dead, but the survivors are still alive and will begin again. So the 'open' end of the horseshoe points to the Solstice Sunrise, but the 'main' direction of the symbol points in the opposite direction – the winter solstice sunset.

Furthermore, there in the centre of the Circle lays a very special stone - the Altar stone. The reason it is special is two-fold. Firstly, it's made of a material unlike the other Sarsen standing stones, it called mica. "Hawkins makes note that while all the other stones were either bluestone or sarsen, the so called altar-stone is 'of fine-grained pale green sandstone, containing so many flakes of mica that its surface, wherever freshly exposed, shows the typical mica glitter'. Currently, geologists are trying to locate the source of this sandstone in Wales. Sadly, they are looking in the wrong direction as I would imagine it's from their homeland and currently under less than 30m of water in Doggerland.

But this is not the only piece of Sarsen that is made of 'mica-sandstone'. The most important piece of this special mica-sandstone lays by the moat of the monument to the north east and is called the 'Slaughter Stone'. I will not delve into the reason that this stone is called the Slaughter Stone, but it is interesting that even the druids some 4000 years after the stone was laid into the ground still associated it with death.

Most archaeologists believe that the slaughter stone was once a standing stone at the entrance of the monument, this flawed ideas is the result of a hypothetical drawing by Inigo Jones in 1655. This drawing shows Stonehenge as a perfect circle with hexagon trilithon interior and three entrances into the site with six erect standing stones as access points, of which the Slaughter Stone was one. This idea was incorporated in John Aubrey's drawing in 1666, which was more accurate, but again had the tendency to place all the fallen stones in upright positions.

This false assumption was further compounded by William Cunning in 1880, when (it was reported) that he suggested his grandfather "saw" the upright slaughter stone in the 17th Century). This mistake was latter corrected yet the myths amongst archaeologists still remain (Stones of Slaughter, E Herbert Stone, 1924 pp120).

timeline

4500 BCE	2500 BCE	800 BCE	0-400 AD	1-2000 AD
Neolithic Age Begins	Bronze Age Begins	Iron Age Begins	Roman Period	Written History

The reality is that the Slaughter Stone was always (like the Altar Stone) a deliberate recumbent as the excavations of this stone by Hawley and Newall in the 1920's clearly show. As the chalk subsoil was also deliberately flattened before it was placed in its current position, Hawley presumed that the Slaughter Stone was once 'buried',. This idea is understandable as the stone does lay below ground level, but what Hawley never understood is that the reason the stone was in this position was for the same reason the ditch was built around Stonehenge, as it was made to be full of water.

Slaughter Stone

This can be observed by the size of the stone hole called 'E' (WA1165) which lays two metres north west of the Slaughter Stone, but still within the 'hollow' that also contains the stone. Most stone holes at Stonehenge are quite shallow – less than a metre in depth, but stone hole 'E' is twice as deep, over 2m. If the Slaughter Stone was placed in it (as some have suggested) it would only be 3m high on the surface, compared to 4.57m for the Heel Stone a few yards away. The only other places in Stonehenge with these large holes is found in the Ditch surrounding the site, which are deep as we have shown to reach the ground water level at construction.

In the past when the Slaughter Stone was placed in this ditch, like the moat, water would have surround the stone similar to an island - for what we see today at Stonehenge is a 6,000 year old model of the land we call Atlantis that lay in the current North Sea.

Not only did they place a piece of special mica-sandstone in a watery ditch surrounded by water, but they also carved out the contours of the island, showing high and low ground like a contour release. Archaeologists have always believed these features were 'weather worn' by age (although the other recumbent stones have not weathered in the same fashion) but recent laser technology has confirmed our belief that this stone was carved as the markings from the tools used can still be seen at microscopic levels.

Finally, and more importantly, the stone has been placed in a very strange position, almost in the way of The Avenue. This shows that this stone and

Cross section of the Slaughter Stone
Showing it was cut into the chalk

17000 BCE

10000 BCE

Ice Age Ends

Mesolithic Age Begins

the Avenue have a connection, this type of connection we see in association with Egyptian Pyramids when 'slight line' are cut into the sides of the burial chambers to important star constellations to show their associations with the gods. At Stonehenge, the Slaughter Stone and the Avenue are important as they link rebirth with the death of this Great Civilisation, for if we look from the Major Sight line stone - the Altar Stone and look towards the Slaughter Stone, it is quite remarkably a direct line POINTING TOWARDS the location of Atlantis in the North Sea, a place we now call Doggerland.

Yet, when you read the two detailed accounts of Plato's work in 'The Timaeus' and 'Citias' where some of the narrative is repeated, but you get a sense that Plato is talking about two different times of Atlantis' history. The original landscape and people, as we have seen his dialogue are not fixed "and it was possible for travellers OF THAT TIME to cross from it to the other islands", "Many great deluges have taken place during the nine thousand years, for that is the number of years WHICH HAVE ELAPSED since the time of which I am speaking". So are we hearing about the original Atlanteans of 9000 years ago who lived in this green and pleasant land or are we being told of a land that was once great during that time then sunk overnight leaving nothing?

The secret is in the text at the start of the story of Solon and what the Egyptians told him, "As for those genealogies of yours which you just now recounted to us, Solon, they are no better than the tales of children. In the first place you remember a single deluge only, but there were many previous ones; in the next place, you do not know that there formerly dwelt in your land the fairest and noblest race of men which ever lived, and that you and your whole city are descended from a small seed or remnant of them which survived. And this was unknown to you, because, for many generations, the survivors of that destruction died, leaving no written word."

What the priests was trying to tell Solon, is that the deluge remembered that Greek history was the one that wiped out Athens and that many such deluges happened in the past - so we can see the confusion with the dates of these floods. Again we see this two-fold story unveiling with Atlanteans referred to as "in your land dwelt the fairest and noblest race of men which ever lived". Does this give us a description of an Atlantean?

We know from DNA the Scandinavians are descended from blonde and blue eyed ancestors with a pale complexion – this description seems to me to confirm that fact, as if we use the word fair, we generally are referring to their hair and features. Moreover, it also shows that two stories are being recited here as within the same book the Atlanteans are also called "various tribes of Barbarians", "a mighty power unprovoked" and "invaders".

Moreover, does the line "And this was unknown to you, because, for many generations, the survivors of that destruction died, leaving no written word." Suggest this race had no written language? It is well known that the Nordic races relied on verbal story telling for their histories. I believe this confirms a 'Nordic' Atlantean. But, the most explosive revelation not picked up by most scholars is the line "and that you and your whole city are descended from a small seed or remnant of them which survived." The priests are telling Solon that the Greeks are 'seeds' of their great nation. Are they implying that they were the first people to colonise Athens and leave their philosophy and beliefs?

The writing then goes on to say "Solon marvelled at his words, and earnestly requested the priests to inform him exactly and in order about these former citizens" and it continues to reinforce this suggestion by saying "You are welcome to hear about them, Solon, said the priest, both for your own sake and for that of your city, and above all, for the sake of the goddess who is the common patron and parent and educator of both our cities." So we now, not only have the Atlanteans 'seeding' the Greek population but also the Egyptian population a thousand years later. The only question left to answer is when, and that is revealed just a paragraph later when he writes "And the duration of our civilisation as set down in our sacred writings is 8000 years. As touching your citizens of nine thousand years ago". So we have our answer, probably the most astonishing revelation in the history of world civilisation, the Atlanteans (" from a distant point in the Atlantic ocean was insolvently advancing to attack the whole of Europe and Asia, to boot" in boats from a land beyond the "pillars of Hercules") seeded and established the kingdoms of Greece nine thousand years ago and Egypt a thousand years later.

The interesting aspect of this two-fold story is that some scholars get confused and believe that Atlantis sunk 9000 years ago, when clearly Plato was talking about the colonisation of Greece and Egypt, and then goes on to talk about the Atlanteans coming to conquer and enslave the Greeks and Egyptians, which is clearly a later date. When he does talk about Atlantis sinking, no dates are offered. This is when the second story totally conflicts with the first, initially they were "the fairest and noblest race of men which ever lived" but within a paragraph or two become "For these histories tell of a mighty power which unprovoked made an expedition against the whole of Europe and Asia, and to which your city put an end. This power came forth out of the Atlantic Ocean, for in those days the Atlantic was navigable; and there was an island situated in front of the straits which are by you called the Pillars of Hercules.

17000 *BCE*   **10000** *BCE*   You are HERE

Ice Age Ends   Mesolithic Age Begins

This vast power, gathered into one, endeavoured to subdue at a blow our country and yours and the whole of the region within the straits; and then, Solon, your country shone forth, in the excellence of her virtue and strength, among all mankind. She was pre-eminent in courage and military skill, and was the leader of the Hellenes. And when the rest fell off from her, being compelled to stand alone, after having undergone the very extremity of danger, she defeated and triumphed over the invaders, and preserved from slavery those who were not yet subjugated, and generously liberated all the rest of us who dwell within the pillars."

This dichotomy does not make sense unless we are talking about two separate time periods. Nine thousand years ago Athens was being seeded and Egypt did even exist as a civilisation for another one thousand years. But the Atlantean army set forth to conquer and enslave, but was defeated by a nation that was in its infancy! Clearly, this shows two stories with independent time lines. With this in mind, we can date the end of Atlantis after the armed conflict and deluge that destroyed Athens. For Plato writes "But afterwards there occurred violent earthquakes and floods; and in a single day and night of misfortune all your warlike men in a body sank into the earth, and the island of Atlantis in like manner disappeared in the depths of the sea. For which reason the sea in those parts is impassable and impenetrable, because there is a shoal of mud in the way; and this was caused by the subsidence of the island." B.Jowett (1871)

But another interpretation is "But at a later time there occurred portentous earthquakes and floods, and one grievous day and night befell them, when the whole body of your warriors was swallowed up by the earth, and the island of Atlantis in like manner was swallowed up by the sea and vanished; wherefore also the ocean at that spot has now become impassable and unsearchable, being blocked up by the shoal mud which the island created as it settled down". W.R.M.Lamb (1925)

I think the key here is 'But afterwards' or 'But at a later time', this was a time period unknown 'after' the conflict and I would imagine that conflict would be many thousands of years after the establishment of Greece and Egypt with independent armies and leadership. We know from geologists that the last piece of Doggerland sank in about 4000 BC, which is about the time that the Greece Empire was at its height and the Atlantis Empire was all but gone, so it's my opinion this is the time period Plato was referring too. This would also make sense of the line "For which reason the sea in those parts is impassable and impenetrable, because there is a shoal of mud in the way; and this was caused by the subsidence of the island." For even today some 6,000 years later, we refer to the area as the 'dogger sand bank'.

Finally, there is evidence within Plato's writing that there was contact with the Atlanteans even after the flood had taken way their major city Atlantis, for a line in Critias talking about the final flood gives us get

4500 BCE	2500 BCE	800 BCE	0-400 AD	**timeline** 1-2000 AD
Neolithic Age Begins	Bronze Age Begins	Iron Age Begins	Roman Period	Written History

another clue to the location of this legendry land, for Plato writes " there are remaining in small islets only the bones of the wasted body, as they may be called, all the richer and softer parts of the soil having fallen away, and the mere skeletons of the country being left." So what was left after Doggerland disappeared? We have to remember this sea level increase was happening all over the lost continent, so much so the Irish Sea separated Ireland from Wales, the English Channel was formed separating Britain from mainland Europe and the North Sea took the last of the peninsulas land mass and island to become a vast watery landscape – but the British Isles remained. In comparison would you not call that the 'bones' of what was formerly there, and does the makeup of this island look like a 'skeleton of a country'?

Map by Abraham Ortelius, Amsterdam 1572: at the top left Oceanvs Hyperborevs separates Iceland from Greenland

Hyperborea

It would be wrong to imagine that Plato's writing's are the only one that mentions an ancient advanced civilisation that lived in prehistory, although he is the only one to mention Atlantis by name. The Greek scholar Herodotus (Histories, Book IV, Chapters 32-36) some 500 years BEFORE Plato talks about a myth of an ancient land called Hyperborea. Remembering that the Atlanteans had 'seeded' this culture in about 9000 BCE such myths should be considered, if not as solid as Plato's works, but just as valid to endorse the location and description of the land, for the more pieces of the jigsaw we find the clear the picture we will obtain.

The Hecataeus of Abdera in the 4th Century BC, believed Hyperborea was in Britain as:

"In the regions beyond the land of the Celts there lies in the ocean an island no smaller than Sicily. This island, is situated in the north and is habited by Hyperboreans, who are called by that name because their home is beyond the point whence the north wind blows; and the island is both fertile and productive of every crop, and has an unusually temperate climate"

From Diodorus Siculus (ii.47.1-2). Also the sun was supposed to rise and set only once a year – as it does in the North Eastern area of Atlantis known today as northern Sweden and Norway. Hecateaus of Abdera also wrote that the hyperboreans had a 'circular temple' on their island – is this Stonehenge or the temple of Apollo on the Plain of Atlantis?

17000 BCE   **10000** BCE   You are HERE

Ice Age Ends   Mesolithic Age Begins

Gothicism (is the name given to what is considered to have been a cultural movement in Sweden, centred around the belief in the glory of the Swedish ancestors, originally considered to be the Geats, which were identified with the Goths) in the 17th century in Sweden, declared the Scandinavian peninsula was both Atlantis and Hyperborean land. Ptomolemy (Geographia, 2.21) and Marcian of Heraclea (perplus, 2.42) both placed Hyperborea in the north sea which they called the 'Hyperborean Ocean'.

In conclusion, Hyperborea was a fabulous realm of eternal spring located in the far north beyond the land of winter. Its people were tall and blessed with , long-lived race free of war, hard toil, and the ravages of old age and disease. The land is usually described as a continent-bound land, bordered by the great earth-encircling river to the north (The Atlantic), and the great peaks of the mythical Rhipaion mountains to the south (The Alps). Its main river was the Eridanos (The Danube), which flowed south, drawing its waters directly from the Okean-stream (The Norwegian Trench). The shores of this stream were lined by amber-bearing poplar trees (as the Baltic Sea is at its source) and its waters inhabited by flocks of white swans. Blessed with eternal spring, the land producing two crops of grain per year. But most of the country was wild, covered with rich and beautiful forests, "the Garden of Apollo".

The Golden Age

According to some modern thinkers (Blavatsky, Guenon and Evola), Hyperborea was the terrestrial and celestial beginning of civilisation. The home of original Man. Some theories postulate Hyperborea was the original 'Garden of Eden', the point where the earthly and heavenly planes meet. And it is said Man transgressed Divine Law in this 'Golden Age' civilisation, the ultimate price being his banishment to the outside world. Man ventured into other regions of Earth, establishing new civilisations, bringing to an end this great and glorious Golden Age.

The Golden Age is central to manifold ancient traditions and myths. Significantly, the Golden Age appears most frequent in the traditions of cultures stretching from India to Northern Europe — the area directly beneath the Polar regions. Joscelyn Godwin, in Arktos, The Polar Myth in Science, Symbolism and Nazi Survival, says:

"The memory or imagination of a Golden Age seems to be a particularity of the cultures that cover the area from India to Northern Europe... But in the ancient Middle East there is an obvious relic of the Golden Age in Genesis, as the Garden of Eden where humanity walked with the gods before the Fall. The Egyptians spoke of past epochs ruled by god-kings. Babylonian mythology... had a scheme of three ages, each lasting while the vernal [Spring] equinox precessed through four signs of the zodiac; the first of these, under the dominion of Anu, as a Golden Age, ended by the Flood. The Iranian Avesta texts tell of the thousand-year Golden Reign of Yima, the first man and the first king, under whose rule cold and heat, old age, death and sickness were unknown."

timeline

4500 BCE	2500 BCE	800 BCE	0-400 AD	1-2000 AD
Neolithic Age Begins	Bronze Age Begins	Iron Age Begins	Roman Period	Written History

The memory of a Golden Age, although rendered in an archetypal or mythological form, serves a super-historical purpose. This is why the remembrance of the ancient civilisation of Atlantis is sometimes enmeshed with that of Hyperborea. All myths are known to have a historical basis. Transmitted primarily by oral tradition, they are wrapped in a catchy and simple tale that ensures their survival and transmittal down through the ages. Myth serves an extremely vital function — a recollection of our beginnings, knowledge of where we are heading, and what we are supposed to do.

This I believe is the final piece of evidence that proves that Doggerland and the NEP was the continent and peninsula that is referred in history as Atlantis and we can now look at the monument they built to their land and see just how advanced this civilisation really was - and this is what we will do in the next two books in our epic trilogy.

17000 BCE   10000 BCE

Ice Age Ends   Mesolithic Age Begins

Appendix A

Listing of hypothesis proofs, with page references.

Geological

Proof of Hypothesis No.1 - due to the size and location of the last glaciation, water from the ice cap must have flooded the landscape causing the rivers to rise and turning the landscape into a 'flooded island environment' (Page 20).

Proof of Hypothesis No.2 - water from the ice cap from the last Ice Age flooded the British landscape resulting in newly formed and enlarged rivers with islands – this groundwater slowly receded from the land and moved to the North and Irish Seas, creating the English Channel and flooding Doggerland (Page 26).

Proof of Hypothesis No.3 - the Isostatic transformation of the landmass during the last ice age that left the landmass nearly half a mile below the sea level would have raised the level of the WATER TABLE relative to the landmass (Page 33).

Proof of Hypothesis No.4 - the River Thames was ten times larger in the Mesolithic period than today and of 'fresh water' as the sea level was 30m lower. This river had to be feed by other rivers to obtain the necessary volume to exist. Therefore, the rivers that feed the Thames were also ten times larger than the same rivers today. This could only happen if the ground water levels were consequently higher (Page 37).

Proof of Hypothesis No.5 - sedimentary deposits from the South Downs, geologists believe to be from ancient rivers created during the last ice age having now been carbon dated to 4290 BCE. This indicates that these waterways were still active (wet) 7,000 years ago and not 17,000 as geologists have previously assumed (Page 40).

Proof of Hypothesis No.6 - abnormally high peat deposits in Britain, compared to our 'southerly' neighbouring countries bears testament that the British landscape was in recent times (last 9,000 years) must have been flooded for an extensive period (Page 46).

Proof of Hypothesis No. 7 - The Somerset flats are well known to have flooded in the recent past because of sea intrusion. but during the Mesolithic Period, the sea level was 5-35m lower than today. The 100+ islands reported by Dr Brunning can only exist if the water was fresh water from local rivers, due to higher ground water levels (page 49).

Archaeological

Proof of Hypothesis No. 8 – Wansdyke has no military advantage as it seems to end without any defences on both edges. Yet when our proposed prehistoric waterways are introduced, both ends of this earthwork meets shorelines showing that it was in fact a canal (page 55).

timeline

4500 BCE	2500 BCE	800 BCE	0-400 AD	1-2000 AD
Neolithic Age Begins	Bronze Age Begins	Iron Age Begins	Roman Period	Written History

Proof of Hypothesis No. 9 - the Winterbourne Crossroads Long Barrows are built originally on the ancient shorelines from our hypothesis. But as time went by the shoreline receded from these original points, so our ancestors dug 'canals' to join the Long Barrows to the receding shoreline (page 57).

Proof of Hypothesis No. 10 - Hawley found evidence of a 'water sealant' in the bottom of the ditch at Stonehenge, described as a 'layer of foot-trampled mud'. This 'water sealant' consisted of a mixture of clay/mud, chalk and struck flint. This arrangement is only usually found at the bottom of Mesolithic/Neolithic period 'dew ponds' – clear evidence of groundwater at Stonehenge, meaning that the ditch was a moat, not a defensive fortification (Page 60).

Proof of Hypothesis No. 11 - Hawley called it the 'dark layer'. It was the remains of decayed organic matter and sediment – clearly showing that the Stonehenge ditch once was filled with groundwater (Page 61).

Proof of Hypothesis No. 12 - irregular ditches were deliberately cut at different depths and with low level internal walls to allow groundwater to flow over the ridges. (Page 62)

Proof of Hypothesis No.13 - the construction of a moat with a chalk sub-soil shows that the site was deliberately chosen to allow groundwater to freely flow into the moat at high tide – providing not only a groundwater filled feature but pure water, clean enough to drink. (Page 63)

Proof of Hypothesis No. 14 - these smaller moats that feed from the main moat at Stonehenge would only have been constructed if the main moat contained groundwater. (Page 64)

Proof of Hypothesis No.15 - the discovery of 'Bluehenge' by the River Avon proves that the Bluestones were located by water in order to be effective and provide their healing properties. Consequently, it is entirely feasible that Stonehenge, with its famous bluestone circle, must also have been constructed by to a water source. (Page 66)

Proof of Hypothesis No. 16 - the extensive number of Bluestone chippings, in proportion to Sarsen Stone chippings leaves us to concluded that the Bluestones were deliberately broken up to be used in bathing within the moat. (Page 68)

Proof of Hypothesis No. 17 - the post holes in the Stonehenge car park are approximately 1m in diameter and dug in a line, would reflect a shoreline, if the water table was higher. The only reason to construct post holes is to bear weight from above to allow the mass to dissipate within the hole. This proves that the holes were used as mooring stations with a cross piece being used as an ancient crane. (Page 69)

Proof of Hypothesis No. 18 - the increased water tables in Mesolithic Britain would have made it a far easier task to move the Bluestones by boat, via a direct water route from the Preseli Mountains in Wales. (Page 70)

Proof of Hypothesis No. 19 - the post holes found in the car park at Stonehenge have been dated between 8500 BCE to 7500 BCE. These holes lay on the exact position and height of the prehistoric low tide shoreline. This not only proves existence of water but also proves the date of Stonehenge's first construction. (Page 74)

Proof of Hypothesis No. 20 – the piece of Rhyolite/Bluestone discovered in the 1989 excavated post hole, has been dated between 7560 BCE to 7335 BCE, proving not only that the original post hole is older than

17000 BCE   *10000 BCE*

Ice Age Ends   Mesolithic Age Begins

archaeologists believe, and more importantly, offers the first real date of when the Bluestones arrived at Stonehenge and construction began. (Page 75)

Proof of Hypothesis No. 21 - the existence of the R & Q post holes, forming the semi-circle, pointing in a North West direction and both the geophysical and missing post hole evidence, proves that the original monument was orientated towards North West and the mid-summer moon setting. Consequently, the path from the centre of the Bluestone crescent moon leading to the landing site and our hypothesis's Mesolithic shoreline. (Page 77)

Proof of Hypothesis No. 22 - the presence of a palisade to the North West of the monument which joins two predicted shorelines, proves that water was present at the construction stage of Stonehenge and moreover, that it was used an excarnation site. (Page 78)

Proof of Hypothesis No. 23 - the variation in the size of the ditches either side of the Avenue, to allow an even depth of water on both sides, shows that water was present at Stonehenge in Neolithic Times. (Page 81)

Proof of Hypothesis No. 24 - according to 'The Environs Stonehenge Project', the Avenue terminated at the 'elbow' of the processional causeway in an abrupt ending, without reason. Our hypothesis explains that the reason for the termination was the predicted shoreline of Stonehenge during the Neolithic Period. (Page 82)

Proof of Hypothesis No. 25 - excavations undertaken between 1988 and 1990 in the 'elbow' part of the Avenue, show that 14 large post holes were positioned at an angle that would meet the predicted shoreline of my hypothesis during the Neolithic Period. (Page 83)

Proof of Hypothesis No. 26 – the lack of Mollusca and the finding of calcium carbonate, at certain levels in the ditch of Coneybury Henge proves that the moat surrounding the monument was full of water due to the high water table predicted during the Mesolithic Period by my hypothesis. (Page 86)

Topological

Proof of Hypothesis No. 27 – the 50 barrows that surround Stonehenge are all built over 75 m above sea level, but below the tops of the hills upon which they are constructed. Statistically, this would only happen if something prevented building below that level – and that something was water. (Page 90)

Proof of Hypothesis No. 28- the positioning of long barrows clearly indicates that they were built at the shorelines of prehistoric rives, because the traditional paths from existing rivers indicate that the processional causeways would have either passed higher ground or overshot the brow of the hill. (Page 91)

Proof of Hypothesis No. 29 – The eight long barrows that surround Stonehenge were deliberately positioned parallel to the shoreline of the waterways, at no particular compass orientation. Their 'sterns' were deliberately constructed to be seen from a distance, to act as a direction indicator. (Page 93)

Proof of Hypothesis No.30 - The elevation data shows that Stonehenge was placed three-quarters of the way up a hill, which matches the predicted Mesolithic water tables. (Page 99)

timeline

4500 BCE	2500 BCE	800 BCE	0-400 AD	1-2000 AD
Neolithic Age Begins	Bronze Age Begins	Iron Age Begins	Roman Period	Written History

Proof of Hypothesis No. 31 – The Cursus was created as a representation of the journey between life and the after world – on one side was a long barrow, on the other an island to represent their world, and in the middle the water. The line of the sunrise and moonset cuts through the centre of Stonehenge. (Page 109)

Proof of Hypothesis No. 32 – The depressions to the North and South of Old Sarum show that, in Mesolithic times, these features were used as harbours on the island that was Old Sarum. (Page 114)

Proof of Hypothesis No. 33 – The inner ditch at Old Sarum is of a size and depth that would only fill with water if the water table were 30 m higher than today. This height matches the shoreline of our hypothesis. (Page 115)

Proof of Hypothesis No. 34 – The history of Old Sarum shows that the water table dictates the uses and functions of the prehistoric island. If we look at the receding water levels at Old Sarum when Salisbury Cathedral was abandoned, we can reverse engineer the water levels during the Mesolithic Period – they match our hypothesis. (Page 116)

Proof of Hypothesis No. 35 – The long barrows around Avebury are built on the shorelines of the Mesolithic waterways predicted by my hypothesis. Not only are they on the shoreline, but their orientation proves that they were used to navigate ships between Avebury and Stonehenge. (Page 118)

Proof of Hypothesis No. 36 – The ditches at Avebury are over 11 m deep. The excavated chalk was not used to bank the inner side of the ditch as expected if it was a defensive feature – but on the outside to shelter the circle. Therefore, the only possible reason for such an excavation would be to create a moat for boats to harbour. (Page 119)

Proof of Hypothesis No. 37 – The erosion of the chalk banks at Avebury give us an indication the approximate date the monument's ditches were first constructed. At 1mm per annum this indicates the ditches were built BEFORE 6000 BCE (Page 120)

Proof of Hypothesis No. 38 – Silbury Hill was constructed during the Neolithic Period, once the waters had subsided. The location of the hill is at the end of the waterway predicted by my hypothesis. With its close connection to the Sanctuary mooring site, it represents a continued connection between Stonehenge and Avebury. (Page 121)

Proof of Hypothesis No. 39 – Durrington Walls shows clearly that it was originally constructed not as an encampment, but as a harbour. The V-shaped floor and sloping profile is perfect for mooring ships and boats, and the high banks on three sides would give shelter. This also proves that the waterline in the Mesolithic Period was at the height proposed by my hypothesis.(Page 124)

Proof of Hypothesis No. 40 – During excavations in 1966 at Durrington Walls, the post holes of a mooring station were found on the Neolithic shoreline predicted by my hypothesis. (Page 125).

17000 BCE   *10000* BCE

Ice Age Ends   Mesolithic Age Begins

INDEX

timeline

4500 BCE	2500 BCE	800 BCE	0-400 AD	1-2000 AD
Neolithic Age Begins	Bronze Age Begins	Iron Age Begins	Roman Period	Written History

17000 BCE   **10000** BCE

Ice Age Ends   Mesolithic Age Begins

timeline

4500 BCE	2500 BCE	800 BCE	0-400 AD	1-2000 AD
Neolithic Age Begins	Bronze Age Begins	Iron Age Begins	Roman Period	Written History

17000 *BCE*   **10000** *BCE*

Ice Age Ends   Mesolithic Age Begins

4500 BCE	2500 BCE	800 BCE	0-400 AD	1-2000 AD
Neolithic Age Begins	Bronze Age Begins	Iron Age Begins	Roman Period	Written History

Acknowledgements and Disclaimer

References

Page 28[1]   Canadian Encyclopedia: http://www.thecanadianencyclopedia.com/articles/glaciation
Page 32[2]   Macklin and Lewin 2003; Knight and Howard 2005; Greenwood and Smith 2005: Smith et al 2005.
Page 34[3]   Geomorthology 33 (2000) 167-181. D. Maddy et al.
Page 41[4]   Wikipedia: http://en.wikipedia.org/wiki/Geology_of_Great_Britain
Page 45[5]   Wikipedia: http://en.wikipedia.org/wiki/Peat
Page 47[6]   Wikipedia: http://en.wikipedia.org/wiki/Star_Carr
Page 48[7]   Wikipedia: http://en.wikipedia.org/wiki/Somerset_Levels
Page 49[8]   See British Geological Society web site for an animation; http://www.bgs.   ac.uk/discoveringGeology/climateChange/general/ seaLevelChangeCaseStudies.html
Page 50[9]   BBC News   http://www.bbc.co.uk/news/uk-england-somerset-14239742
Page 53[10]   Wikipedia: http://en.wikipedia.org/wiki/Levee
Page 53[11]   Wikipedia: http://en.wikipedia.org/wiki/Offa's_Dyke
Page 55[12]   Wikipedia: http://en.wikipedia.org/wiki/Wansdyke_(earthwork)
Page 57[13]   English Heritage: http://www.pastscape.org.uk/hob.aspx?hob_id=219525
Page 70[14]   Mostly Herbs: http://www.mostlyherbs.com/Antiseptics.html
Page 129[15]   What-When-How: http://what-when-how.com/ancient-europe/brzesc-kujawski- consequences-of-agriculture 5000-2000-b-c-ancient-europe
Page 136[16]   Wikipedia: http://en.wikipedia.org/wiki/Gift_economy

Images

Page 7   The tuscan magazine - www.thetuscanmagazine.com
Page 8   Spread of farming - www.humanpast.net
Page 9   Bronze Age Ireland - www.dayof archaeology.com
Page 10   House MD - fallingdark.deviantart.com
Page 13   The Vitruvian man and Da Vinci - www.utaot.com
Page 14   http://www.freepspwallpaper.org/tag/drama-film
Page 22   Jelgersma (1979)
Page 24   Hydrology - Uk Groundwater forumn
Page 38   Ouse & Adur Rivers Trust - www.oart.org.uk
Page 42   British Geological Society

17000 *BCE*   **10000** *BCE*

Ice Age Ends   Mesolithic Age Begins

Page 43/44	BGS Borehole viewer - http://mapapps.bgs.ac.uk/boreholescans/boreholescans.html
Page 45	JRC Landmanagement - wwweusoil.jrc.ec.europa.eu
Page 48	Somerset Flats - Somerset County Council Archaeological Service
Page 61/	Stonehenge in its landscape - cleal et al 1995
Page 73	Wessex Archaeology
Page 79	The Ancient Art of Enchanting the Landscape - http://thehobgoblin.blogspot.co.uk/2012/07/ritual-pits.html
Page 87	Messa Community College - http://web.mesacc.edu/dept/d10/asb/anthro2003/archy/process/page3.html
Page 77	English Heritage - Report on Geophysical Surveys Sept 2010 - July 2011
Page 92	Iona Miller - http://ionamiller2010.iwarp.com/whats_new_41.html
Page 101	Robin Heaths Blog - http://www.matrixofcreation.co.uk/megalithic-sciences/item/93-moving-the-bluestones-at-last-a-successful-method-is-demonstrated
Page 122	Ancient Celtic |New Zealand - http://www.celticnz.co.nz/DurringtonWalls/Durrington%20Henge%201.htm
Page 131-3	THEORETICAL STRUCTURAL ARCHAEOLOGY - http://structuralarchaeology.blogspot.co.uk
Page 134	Transcotland http://www.transcotland.com/crannog.htm
Page 134	Archaeology in Marlow - http://www.archaeologyinmarlow.org.uk/2011/01/a-star-is-born
Page 141	Doggerland - http://log.doggerland.net/2010/10/21/a-hypothetical-landscape
Page 143	BBC News - http://www.bbc.co.uk/news/uk-scotland-edinburgh-east-fife-18687504
Page 150	ZME Science - http://www.zmescience.com/research/lost-city-of-atlantis-found-discovered-0205021
Page 154	Adam Standfor Photography - http://www.adamstanfordphotography.co.uk/in-the-round-2
Page 156	Myth & Popular Culture - http://mythandpopculture.wordpress.com/2010/11/09/norse-mythology
Page 157	Pillars of Hercules (wikipedea) - http://en.wikipedia.org/wiki/Pillars_of_Hercules
Page 160	Architects Journal - http://www.architectsjournal.co.uk/culture/utopias-sustainism-and-architecture-between-the-possible-and-the-impossible/8611762.article

timeline

4500 BCE	2500 BCE	800 BCE	0-400 AD	1-2000 AD
Neolithic Age Begins	Bronze Age Begins	Iron Age Begins	Roman Period	Written History

Authors Details

This book is the first part of a trilogy 'Prehistoric Britain' and is a result of over 30 years of studying archaeological sites throughout the world.

Robert John Langdon is a writer, historian and social philosopher who has worked as an analyst for the government and some of the largest corporations and education institutes in Britain including British Telecommunications, Cable and Wireless, British Gas and University of London.

Since his retirement five years ago, Robert has studied Archaeology, Philosophy and Quantum Mechanics at University College London, Birkbeck College, The City Literature Institute and Museum of London. He has three children and two grandchildren and lives on the South Downs in East Sussex, where he owns a book shop which sells his literature and associated maps and prints.

He has an active blog site through which he can be contacted and asked questions on his books and hypothesis; *www.the-stonehenge-enigma.info*

Information about the other books in the trilogy can be found at: *www.prehistoric-britain.co.uk*

17000 BCE	*10000 BCE*
Ice Age Ends	Mesolithic Age Begins